Brian Cathcart is author of *The Case of Stephen Lawrence*, *Jill Dando: Her Life and Death* and *Were You Still Up for Portillo?* He was previously deputy editor of the *Independent on Sunday*. He lives in London.

Rain

BRIAN CATHCART

Granta Books
London

Granta Publications,
2/3 Hanover Yard, Noel Road, London N1 8BE

First published in Great Britain by Granta Books 2002

Copyright © 2002 by Brian Cathcart
Illustrations copyright © 2002 by Dave Hopkins

The permissions that appear on page 101 constitute an
extension of this copyright page.

A CIP catalogue record for this book is available
from the British Library.

1 3 5 7 9 10 8 6 4 2

Typeset by M Rules

Printed and bound in Great Britain by
Mackays of Chatham PLC

To Patrick

Into each life some rain must fall.

HENRY WADSWORTH LONGFELLOW

stratocumulus

1

In the six hundredth year of Noah's life, in the second month, the seventeenth day of the month, the same day were all the fountains of the great deep broken up, and the windows of heaven were opened.

Genesis 7:11

Right now, as I sit at my desk, gusts of wind are whipping the rain against the window beside me. The big drops hit the panes and streak in diagonal lines across the glass, breaking into little pebbles of light. They settle, sag under their own weight and then zigzag fitfully to the bottom of the frame. Another gust and other drops take their place – hundreds, perhaps thousands of drops hurled against a window the size of a big table. Is there a word to describe that noise? It's like fine gravel being thrown against the glass except that it is definitely, audibly wet – a white sound made up of millions of tiny splashes. Out beyond the windowpane the view stretches over some garage roofs to a curtain of tall trees where I can see branches thrashing in the wind. The colour has been bleached from the scene to the point where it is almost monochrome, for this is heavy rain, regimented armies of grey drops slashing through the air and billowing around the angles of the building I live in. I can't describe the cloud from which all this is coming because I can't see it – when I look up all I see is the rain falling. It is three in the afternoon and not yet winter and I have the light on.

In the past few years I have watched a lot of rain through my big window. Sometimes it has been light, what my mother used to call 'spitting' and the meteorologists know as drizzle, but often it is heavy like this and on a couple of dozen occasions it has been awe-inspiring, coming down from a thunderous sky in straight lines like iron bars that no wind could bend. On those days I have found myself wondering how the air could possibly support so much water. How could it ever have lifted up such an ocean of liquid and brought it to that space over our heads? Think of the weight of a bucket of water; this must be millions of buckets.

Of course it doesn't rain like this every day. We have dry weeks, even relatively dry months, but overall the years since 1997 or so have been wet in a way I do not remember experiencing before. The rain seems to have become more persistent, more enduring, and once I began to take notice I kept coming across little signs of it. Here is one: my older son is twelve and from the time he was seven he used to play football with his friends in their local London park on Sunday mornings. Two years ago, however, the games came to a halt, not because the boys grew out of football – they still play it often – but because the ground became waterlogged. Writing at the end of 2001, I would say they have played in that park only a couple of times since the autumn of 1999. Often it was raining, but even when it was not, the mud and wet grass made football in the park an unappealing prospect. At one stage a pond formed over the goal area at one end and it remained there so long that a couple of ducks took up residence. Now the boys never think of playing there.

Here is another sign: I noticed recently that I hardly ever wear my leather jacket any more, even though for two or three winters I wore it almost every day, and I realize that I have abandoned it because it is useless in the rain, soaking up water until it weighs a ton. The jacket keeps me warm, but keeping warm has not been a problem; rain has. And here's another: I have just watched a television programme in which a historian toured the country recounting episodes from Britain's military past, and what struck me most was that everywhere he popped up was extremely wet, from the squelching military road in the Scottish Highlands to the waterlogged defences of Winchelsea in Kent. As if to emphasize this, he never appeared on screen without a huge waxed coat draped over him.

The newspapers seem to confirm the impression. Open a paper any time between September and April and there will usually be a story of flooding or flood warnings somewhere in the country. Floods are so frequent that reporters long ago ran out of original language to describe them and the television footage of a calamity in, say, Essex, is indistinguishable from another months earlier in Gloucester. Then there are those weather records that have made headlines. Not long ago we lived through the wettest twelve-month period ever recorded in England and Wales. It ran from the spring of 2000 to the spring of 2001 and it included a number of lesser highlights – for example April 2000, with well over double the average rainfall, was the wettest April on record, while September, October and November 2000 combined to make the rainiest autumn. The period since then may not have been quite so remarkable but by and large it has been wet too, so that the four years 1998–2001 saw heavy rainfall levels of a sustained kind only rarely seen in the recorded past, while 2002 began in much the same way, with heavy rain and floods in

February. The headlines also bring us a new and watery view of global warming: once we were told to expect vineyards spreading north through England and restaurants spilling out on to every pavement – but now the forecast for the twenty-first century is rainy. It is a good moment, then, to stop and think about rain.

2

It is impossible to live in a country which is continually under hatches . . . Rain! Rain! Rain!

John Keats, in a letter to J. H. Reynolds (1818)

In 1935 Robert Watson-Watt, the inventor of radar, wrote a popular book on meteorology called *Through The Weather House*, which contains this curious observation:

> Civilization, whatever the dictionaries may say, means getting above the weather . . . the oldest mother of invention was the need for devising ways of doing what we want to do, in spite of the weather.

The same idea, viewed from a different perspective, crops up in an early Tom Stoppard play, *Enter a Free Man*, when a pub philosopher remarks:

> I don't know if you've ever thought, George, but if you took away everything in the world that had to be invented, there'd be nothing left except a lot of people getting rained on.

This is not the usual way to measure civilization – things like health, equality and intellectual attainment spring more readily to mind – and yet it makes some sense. The need to escape the cold and the wet, which played its part in the building of the first houses and in the stoking of fires in the earliest hearths, has always had a strong influence on human beings and still does. Damp-proof courses, Gore-Tex, package holidays and sports stadiums with roofs all bear witness to it today, while on a grander scale the modern hegemony of the motor car reflects at least in part that imperative to 'do what we want to do in spite of the weather'. What is surprising, however, is how recently the real advances have been made, because for most of the long existence of our species in this part of the world life has been not only short and often brutish, but also distinctly damp. Two centuries, in fact, are enough to tell the story.

In the years around 1800 half a dozen men were alive who would contribute in important, almost revolutionary ways to our contest with the rain. One of them was John Loudon McAdam, a Scot who might easily have become an American. Born in 1756 to a well-off family in Ayrshire, western Scotland, he was sent at the age of fourteen to live with an uncle in New York. Actively loyal to the British in the revolutionary war – he made a living trading in ships and cargoes captured by the Royal Navy – he found himself unwelcome in the new United States in 1783 and so returned to the land of his birth. He bought some land and a share in a factory producing tar for sealing the bottoms of ships, and in due course he moved to the south-west of England. There he put to use an interest he had developed in road-building.

At the end of the eighteenth century British techniques of road-making had not yet recovered to the standards set by the Romans in the first century AD. Rutted tracks, easily transformed to swamps by a shower of rain, were the rule, so that many stretches of road were effectively closed in the rainy winter months and, where they were open, journeys took far longer than in the summer. The effect of a few days' rain could be drastic. In 1736 Lord Hervey wrote indignantly from Kensington:

> The road between this place and London [a distance of two miles] is grown so infamously bad that we live here in the same solitude as if cast on a rock in the middle of the ocean. All Londoners tell us that between them and us is an impassable gulf of mud.

Fifty years later the traveller Arthur Young warned his readers to avoid 'as they would the devil' a particular road in Lancashire:

> They will meet here with ruts, which I actually measured, *four feet deep*, and floating with mud only from a wet summer. What, therefore, must it be after a winter?

Over the centuries much had been said and done about this state of affairs, but to little effect. The methods of financing and administering the roads had been altered and regulations had been placed on the traffic which used them, but the matter of how they were made had been largely ignored. In 1801, with a view to changing that, John McAdam installed himself as surveyor to the Bristol turnpike trustees and took charge of a stretch of west of England road.

Having experimented in road-building on a small scale in Scotland and Cornwall he now perfected his technique, which was both economical and radical. Received wisdom, going all the way back to the Romans, dictated that a good road needed deep foundations of large stones, with graded layers of smaller stones on top. McAdam rejected this, insisting that trenches and heavy foundations were useless, and that what was most needed was to build up the base of a road with earth so that it stood above the level of the surrounding land. 'Having secured the soil from *under* [or underlying] water,' he wrote, 'the roadmaker is next to secure it from the rain water by a solid road, made of clean, dry stone or flint, so selected, prepared and laid as to be perfectly impervious to water.' There would be no big stones at all, only a deep platform of small, evenly-sized, angular ones little larger than gravel. These were given time to settle and the surface was repeatedly levelled until McAdam was happy, after which the road was opened and the action of passing wheels soon created an even, dense, powdery top that was largely impervious to rain. This method, based on what he called 'the application of scientific principles', was so effective that it was to transform travel. No longer did roads need to close in wet weather and no longer did the coach journey from London to Oxford take twice as long in winter as in summer. On a 'macadamized' surface coach services could run in rain or shine. McAdam and his family were soon managing nearly a hundred turnpikes and in due course Parliament ordered his techniques to be applied to the streets of London. (It is a point for pedants, by the way, that his method did not employ tar; 'Tar McAdam' was a personal nickname acquired through his ownership of the ship's tar business in Ayrshire.) John McAdam died in 1836 never having made the fortune which was his due; his technique had proved so useful and so overdue that, though he had several patents, they were generally ignored.

There is nothing to suggest that McAdam ever met his fellow Scot and contemporary Charles Macintosh, although the idea of a conversation between the two – both, in a strange way, in the same line of business – is appealing. Macintosh was born in Glasgow in 1766, the son of a Sutherland mill-owner, and from early youth had a fascination for chemistry. As soon as he was old enough he set up in business, making and supplying a variety of chemicals for industry while conducting experiments on his own account. He was a very good chemist – in time he would be elected a Fellow of the Royal Society – and he made a number of useful discoveries, such as a technique for making bleaching powder and a new method of steel manufacture. Eventually his attention turned to the waste products of the new coal gas industry, which he felt might be turned to good use and profit, and he learned that one of these by-products, coal tar naphtha, had the unexpected property of dissolving easily with rubber. This was a mixture, he saw, that had considerable potential.

At this time there was little in the way of waterproof clothing, the best available being oiled fabrics which were troublesome and offered only limited resistance to rain. Rubber had been around for centuries – since the Spanish conquests in America – and many attempts had been made to incorporate it into cloth, but none met with any success. From 1819 on, however, Macintosh conducted a series of experiments with his new naphtha–rubber solution and in due course he found a way of fixing it between sheets of wool as a sort of sandwich. This new material proved stable and reasonably workable, and above all water ran clean off it, so in 1823 Macintosh was able to patent the world's first practical waterproof fabric. It was a great success and before long the new firm of Macintosh & Co. was turning it out by the ton from a factory in Manchester. Popular demand was boosted by some notable commissions: the Duke of York ordered a blue riding cape in the new material; Guards officers wore it on field exercises; and two Arctic expeditions took sheets of it with them. (There is another pedant's point here: Macintosh himself did not design a coat, so the 'mac' as we know it today was not his creation.) Naturally there were hitches – it was some time, for example, before clothes-makers accepted that seams had to be sealed as well as stitched if they were not to leak – but there was no doubting that the new material was a great liberator. A man who had a Macintosh cloak could ride through a rain shower and emerge with his suit of clothes bone dry.

How far men like Macintosh and McAdam changed life for their contemporaries may be measured in the way that their names have endured. Charles Goodyear, the American rubber pioneer, is another, as is the Italian Luigi Galvani, whose electrical processes revolutionized rustproofing. Sadly no one person can be credited with inventing the umbrella, but it was another liberator. First imported to Britain as parasols – Mary Queen of Scots had one to 'make shadow' – umbrellas were occasionally employed to protect against rain in the eighteenth century, when well-endowed churches sometimes had them to shelter ladies to and from their carriages, and butlers in big houses kept them by the door for the same purpose. However, if a gentleman of that time ventured forth holding one aloft he ran the risk of being thought either unmanly or too poor to afford a carriage. In the early years of the nineteenth century umbrellas were still sufficiently unfamiliar to confuse Sir Robert Peel.

On seeing a fellow traveller holding one furled under his arm during a ferry crossing, Peel addressed him with the words: 'You have a musical instrument there, sir? Might we ask you to favour us with a tune?' Soon lighter materials and better mechanisms brought them into more widespread use and, having been a badge of poverty, they became an indispensable class accessory, even a mark of prosperity – as Robert Louis Stevenson observed: 'It is not everyone that can expose twenty-six shillings' worth of property to so many chances of loss and theft.'

Inventions and developments such as these, taken together with many others whose applications were more general, such as the railways and the electric telegraph, greatly increased people's ability to do what they wanted, whatever the weather. It is easy today to underestimate the degree to which rain, mud and damp had previously been a barrier to communication and economic development. Rain was more than an inconvenience, after all; it was a threat to health. Colds and chills were things to be feared and the idea that they were directly linked with getting wet was well established. In the year 1800 the most popular medical book on the market was still William Buchan's *Domestic Medicine*, first published in 1769, and it was forthright on the subject:

> The most robust constitution is not proof against the danger arising from wet clothes; they daily occasion fevers, rheumatisms and other fatal disorders, even in the young and healthy.

Anybody caught in the rain should change their clothes as soon as possible, Buchan wrote, and he drew a

stern moral from the occasional habit of country people
of sleeping in wet clothes in the open.

> The frequent instances which we have of the
> fatal effects of this conduct ought certainly to
> deter all from being guilty of it . . . Even wet feet
> often occasion fatal diseases.

Jane Austen knew this. 'Getting wet through' on the
ride to Netherfield was enough to land Jane Bennet in
bed for days in *Pride and Prejudice*, while in *Sense and
Sensibility* when Marianne wandered off in the rain in a
moment of romantic despair it all but killed her. To be
dry, to be out of the rain and yet to be able to go about
one's business, was something valuable, even precious.

cumulus congestus

3

Probably the wettest inhabited places in the British Isles are Pny-Gwrhyd Hotel (Snowdonia), Seathwaite Farm (Borrowdale, Cumbria) and Kinlochquoich Lodge (Highland), each with an average of close to 3,120mm of precipitation per year.

from *Climates of the British Isles*, edited by
Mike Hulme and Elaine Barrow

Besides their practical inventions, the people of the early nineteenth century were asserting themselves over rain in another way which was at least as important: they were looking for intellectual mastery. By understanding the rain – what caused it, where it came from and where it went – they might be better placed to improve their defences still further, robbing it of power by robbing it of mystery. This process of discovery proved to be the work of many minds over many decades, but it so happens that some of the most important insights came along around the year 1800.

Unlike McAdam and Macintosh, the name of Luke Howard has not survived as a popular label, and yet this London Quaker did something so bold and ingenious, and at the same time so remarkably obvious, that he deserves to be remembered every time we draw back our curtains in the morning. Howard was born in 1772, the son of a successful manufacturer of iron and tin goods. At the age of fourteen he was sent as an apprentice to a retail pharmacist in Stockport, near Manchester, and there he developed a general interest in the sciences. Returning to London as a grown man he went into business on his own account as a retailer and manufacturer of chemicals and this, so the story goes, entailed his spending a good deal of time walking between the various sites of his operations around the city. Whether it was in the course of these walks or not, he found himself looking at and thinking about the clouds he saw overhead. He noticed their varied shapes and apparent textures, their different heights and colours, the speeds at which they moved and their different propensities to bring rain. Having already dabbled in botany, Howard was familiar with the Latin naming system of Linnaeus, and it occurred to him that clouds might be classified in a similar way. In a lecture given in 1802 he proposed just such a system, later publishing it in an essay that became famous in its day: *On the Modifications of Clouds*. There were, he suggested, three basic types: the high, wispy ones, which he called *cirrus*; the thick, cauliflower ones, which he called *cumulus*; and the drab, low blankets, which he called *stratus*. Between these existed several other varieties, of which the most familiar was that relative of the *cumulus*, the standard rain cloud or *nimbus*. Not only was this idea a brilliant one, but Howard had got it right first time; his classification system was swiftly taken up and, somewhat expanded, it remains in use around the world today. The man himself was taken up too: Constable, that quintessential painter of clouds, made use of his work, while Goethe wrote to Howard in admiration.

With this short step a line was crossed in man's relationship with the weather. As Howard himself pointed out, seafarers and farmers had always studied the sky and many of them had learned to draw inferences from what they saw, but their expertise had always gone to waste. At worst it died with them and at best it found its way into doggerel rhymes and popular sayings which were usually unreliable outside their original context. What had been lacking was what Howard called 'the key of analysis', in this case a system of classification that would allow observers anywhere to record and exchange information of consistent character, so that a store of reliable knowledge about clouds could be created. From this might ultimately emerge the laws of cloud behaviour, and if that could be achieved man need never be mystified or frightened by what he saw in the sky again, indeed he might even anticipate what would happen next. These hopes, to a great degree, were realized, and the same approach would in due course shed light on other aspects of the climate, not least of them rainfall.

Before that, however, there were other matters to be addressed by scientists and the most pressing of these was the question of what caused the rain, for even Howard did not have the answer to that. As with roads, so with the theory of rain: from antiquity to the beginning of the nineteenth century progress had been very limited. The water cycle known to every modern primary school child – evaporation from the sea makes clouds, clouds make rain, rain makes rivers and rivers run to the sea – was unknown to Aristotle, Aquinas and Galileo, and even Newton, who explained so much of our visible world, saw it only through a glass, darkly. From classical times to the Middle Ages men had relied on the notion of the four elements, earth, air, fire and water; provided these elements obeyed certain rules, it was thought, they could simply change, one into another. So, when water in an open container disappeared, it was changing into air, and by the same token when springs bubbled up from the ground, that was earth changing into water. Though the Renaissance put paid to this sort of thinking, it offered no coherent alternative, and the result was centuries of confusion.

Take the case of evaporation. The most casual observation, of water in a bowl or of clothes on a washing line, tells us that water is capable of vanishing into the air at temperatures considerably lower than boiling point. When the scientists of the 1600s and 1700s looked into this they were quick to infer a relationship with the heat of the sun, for they found that water disappeared more rapidly at higher temperatures, but other findings complicated matters. If a cup of water or wet washing were left out overnight, when the sun was absent, evaporation took place nonetheless, albeit at a slower rate. It would even happen on cold nights – in fact, bafflingly, the chemist Robert Boyle established in the 1650s that evaporation could occur *directly from ice* in freezing temperatures and darkness, without any intervening liquid stage. What could this mean?

There was no shortage of theories of evaporation. One was that invisible particles of fire – still thought of as a kind of substance and known as 'caloric' or 'igneous fluid' – penetrated or attached themselves to particles of water and thus provoked the change of state. Another view was that water produced tiny bubbles, or 'vesicles', of rarefied air which naturally floated upwards. The Swiss physicist Horace Bénédict de Saussure announced in 1783 that he had *seen* these vesicles during a climb in the high Alps; he watched through a magnifying glass, he said, as the little bubbles struck a piece of black card and burst open. More persuasive still was the 'solution' theory, which held that water dissolved into air in much the same way that salt dissolved into water. Originating in France, this idea had among its advocates Hugh Hamilton, the professor of philosophy at the University of Dublin and a man of great curiosity and enterprise (later he would attempt to prove the existence of God by scientific experiment). In 1765 Hamilton published *An Essay on the Ascent of Vapours, the Formation of Clouds, Rain and Dew and several other Phaenomena of Air and Water*, which declared that heat could not be the only cause of evaporation, or even the principal one. Water and air, Hamilton argued instead, had a 'mutual attraction' and when they met they had a natural tendency to mix, the one with the other. It was happening in our environment all the time.

> The lowest part of the air being pressed by the weight of the atmosphere against the surface of the water, and constantly rubbing upon it by its motion, has thereby an opportunity of attracting and dissolving those particles with which it is in contact and separating them from the rest of the water . . . The solution of water in air, and the ascent of vapours, is greatly promoted by the motion of the winds, which bring fresh and drier air.

Air, therefore, was like blotting paper which, when it touched water, simply sucked it up and carried it off. For a long time this was the best idea around and it was accepted by Luke Howard, who declared flatly: 'Atmospheric air has an affinity for water.' But the theory had a conspicuous weakness. Since the 1670s scientists had been experimenting with vacuums, observing how natural processes were affected by the absence of air, and they had long ago noticed that in a vacuum evaporation carried on regardless. What this meant, of course, was that air, far from being the vital cause of the process, was not even necessary. At first the theorists bluntly dismissed the results as wrong, but by the early nineteenth century air pumps and laboratory techniques were sufficiently advanced to leave no doubt about the findings. The solution theory, people had to acknowledge, was no solution. Confusion deepened; vesicles were revived and an assortment of electrical theories were produced to fill the gap, but none of them worked either. What was frustrating the scientists, what made nonsense of their calculations and deductions, was the lack of a sufficient understanding of how matter was made up and how its components behaved. This essential knowledge finally made its appearance in the 1850s in the form of the kinetic theory of gases. As is often the way, a number of lonely scientists had glimpsed the big idea long before and been ignored, but now from various corners of the scientific world (credit is due to two more Scots, Robert Brown of Brownian motion fame and James Clerk Maxwell, but it was the work of many hands) the picture began to emerge which is taught in schools to this day. It looks like this.

Water evaporates at temperatures below boiling point because of the behaviour of its molecules. All water molecules, whether in solid, liquid or gas form, have energy: they vibrate and dance. Those in ice move most slowly because they are hooked together to make a crystal, while liquid molecules are also attached to one another, but more loosely – the forces that create the attachments are caused by electrical charges in the molecules. In water vapour, which is what we call the gaseous form of water, the molecules move so fast that they are not affected by these electrical forces; they are free. It is therefore the speed or energy level of the molecules that marks the essential difference between the three states. But those states are not clear-cut; all of the molecules in a mass of liquid water, for example, do not share the same energy level. Some are more active than others and there will always be a minority of them sufficiently energetic to become gas. Where these active molecules are close to the surface in a glass of water, they can break free of their electrical bonds and become separated. This is evaporation. It can happen directly from ice, as Boyle observed, because even in ice some molecules are active enough to jump straight to the gas phase (a process that has its own name: sublimation). It happens in a vacuum, air not being necessary, and, as had long been known, the process accelerates when the temperature rises and slows when it falls.

This important phase in the water cycle was not unravelled, therefore, until the 1850s, and it was some decades later before the explanation was universally accepted. Other phases had to go through a similar process, even the one in which rain turns into rivers. You might think that nothing could be more obvious than this, but Aristotle got it wrong and so, even late in the eighteenth century, did a number of serious British thinkers. They knew what they saw with their eyes – that some rainwater ran off the surface of the earth directly into rivers and that some sank into the ground – but they did not believe that there was enough water in the rain to fill all of the rivers all of the time. Some of it, they speculated, must come from underground lakes or seas, or else it was filtered into the earth in some way from the oceans. In the year 1802, in an early paper delivered in Manchester, these doubts were robustly confronted by John Dalton. Later to become the father of atomic theory, Dalton was a man born and bred in the rainiest corner of England, the Lake District. More or less by rule of thumb he calculated that England and Wales had a 'general mean rainfall' of about thirty-one inches, to which a further five inches could be added for dew. According to the geographers the area of the country was 1,378,586,880,000 square feet, and if the two figures

were multiplied together, Dalton said, the total in cubic
feet would be equal to 115 thousand million tons –
'nearly'. This, therefore, was the weight of water to fall on
England and Wales each year. By employing similarly
brutal methods Dalton arrived at a figure for the amount
of water that passed through the rivers annually and he
declared that, far from being insufficient, the rain was
more than double what was required. There was thus no
need for underground oceans.

So the rain filled the rivers and the rivers flowed into
the sea, where in turn evaporation lifted vapour into the
air. The cycle was taking shape. How and why the water
vapour turned into clouds, however, and how clouds
became rain were still mysteries. Most baffling of all to
the nineteenth-century scientist was the question of
what held the clouds up. It was obvious that they must
be composed of condensed water, which in turn must be
heavier than air, so why did the clouds not just fall
down? Even more provocatively, some clouds existed at
extremely high altitudes where, since the temperatures
were known to be much lower, it seemed certain that
they must be made of ice. When in 1802 the explorer
and scientist Alexander von Humboldt climbed Mount
Chimborazo in Ecuador, then thought to be the highest
peak in the world, he looked up and saw clouds above
his head. And when, two years later, Humboldt's friend
Joseph Gay-Lussac ascended by hot-air balloon to the
almost insane altitude of 6,000 metres, he too was able to
watch clouds above him. How could ice float about up
there? These mysteries were not to be cleared up until
the twentieth century, and before we see how that was
done we must first meet the person who is in many ways
the central figure of this story: George James Symons.

4

St Swithin's Day, if ye do rain
For forty days it shall remain.

> The most enduring of all items of English weather lore, this has no foundation in observed fact. Similar traditions exist in France, Belgium and Germany, although in each country it applies to a different date. St Swithin's Day is 15 July, said to be the day on which, contrary to his deathbed wish, the saint's remains were dug up from their outdoor resting place and transferred inside a church.

Symons was not a man of 1800 but a Victorian; in fact his life falls neatly within Victoria's reign since he was born in 1838, the year after her accession, and he died in 1900, the year before she did. Nor did he come from a particularly rainy region as Dalton and McAdam had done; he was born in London and, barring a few years' schooling in the Midlands, lived in the capital all his life, most of it in the northern district of Camden. Symons grew up at a time when, owing partly to the influence of Luke Howard, meteorology was changing from an enthusiasm for the few to a science for the many, and he clasped it to his bosom at the earliest possible age. As a small boy, therefore, he was fascinated by thunderstorms, devoured the scientific journals and designed his own meteorological instruments, and by the time he was seventeen his interest in the weather was sophisticated enough to earn him election to full membership of the British

Meteorological Society. This learned body, founded just five years previously by a group of worthies and enthusiasts, survives today as the Royal Meteorological Society. Before long Symons was delivering scholarly papers of his own at society meetings and writing thoughtful letters on meteorological matters to *The Times*. And it so happened that his career was blossoming at a time when the British weather was thought to be behaving oddly, for these were years of low rainfall in many parts of the country, even of drought.

In the spring of 1859 a rural correspondent to *The Times* noted the long series of 'deficient' winters and went on:

> Last summer and autumn were remarkable for the scarcity of water in the eastern counties. The labouring poor were paying for it by the pail and the farmers were carting it from a distance into their fields for the use of their stock.

Sir James Glaisher, one of the country's leading meteorologists, was alert to the problem and, drawing on rainfall figures gathered at Greenwich Observatory, gave a measure of these repeated deficiencies of rain. So great were they, he suggested, that questions had to be asked: even if normal rainfall levels returned, would that be sufficient to fill the lakes and rivers again? Worse, was it possible that normal rainfall levels might not return at all? This was no small matter, for the Victorians had reason to be perpetually nervous about water. Not only was theirs the era of railways and heavy industry, it was also the age of sewers and drains, pumps and dams. The cities, with their rapidly growing populations, their thirsty factories and their new water closets, consumed water on a scale never known before and the pressure on the country's rivers and lakes to meet this demand increased steadily. City corporations reached farther and farther afield for their supplies and there was genuine anxiety that in time those supplies might prove insufficient. Glaisher's questions, therefore, were bound to cause alarm.

It was a statistical matter and it had to be admitted that there was only a very poor statistical basis on which to found a discussion. No one, in other words, had a reliable notion of national rainfall patterns. Ever since medieval times there had been occasional, scattered enthusiasts who kept records of the rain in their own areas, but these data had never been collated and studied properly and in any case they lacked consistency of time, place and method. The same was true even of the figures which revealed the 1850s drought; no one could tell how bad it was over the whole country or, with any reliability, how it compared with other years or other periods. A void was revealed: questions were being asked about future rainfall, but if the past and the present were a mystery, what chance was there of understanding the future?

Enter Symons, who now found the mission that was to dominate his life. In 1859 he was twenty-one years old and his father, a London businessman, had died and left him to support his mother. He had a job in trade (it is not clear now what it was) but in his spare time he began to trace by correspondence any amateur weather observers and enthusiasts that he could find. Did they keep daily rainfall figures? If so, could they send them to him? On this basis he began to assemble a set of rainfall figures from around the country for that year. The results, drawing on data supplied by observers in sixty-seven locations, appeared in *The Builder* magazine of April 1860. On their own they did not say much and they certainly did not answer Glaisher's questions, but Symons could already see in what he had done the hint of something far bigger. If such tables were compiled regularly over a number of years it would soon be possible to supply answers to a whole variety of questions about rain which were hitherto more or less unanswerable. These could range from casual enquiries already to be found in the letters columns of the newspapers, such as, 'Has there ever been such a wet day in June?' to the more businesslike, 'What is the worst that the rain can do in this district?' Having had some training as an engineer Symons realized that an essential matter in planning urban drainage was 'the probable maximum fall in the minimum time'.

In 1860 he landed a job at the Meteorological Department of the Board of Trade, which, like the British Meteorological Society, was then something of a novelty. The department had been set up in 1854 under Robert FitzRoy, who had captained the *Beagle* in Charles Darwin's voyage of scientific discovery earlier in his life (and whose name is now commemorated as a sea area in the BBC's shipping forecasts; it replaced Finisterre in 2002). In time the department would become the Meteorological Office ('Met Office'), the country's official forecasting body – though for most of the twentieth century it was a branch of the military establishment. The department should have been the perfect home for Symons but it was not, for he was kept fully occupied with work of FitzRoy's choosing, leaving the gathering and collating of rainfall figures as an occupation for his spare time at home in Camden. Still, in 1861 he was able to publish *English Rainfall 1860*, a four-page pamphlet giving records, not from sixty-seven sites, but now from 168. It proved hugely popular and useful and there followed in later years what might be described as a minor craze for rainfall monitoring and rainfall figures. This was skilfully managed by the young Symons. Recruiting financial assistance from the British Association for the Advancement of Science, he appealed for observers not only through the columns of *The Times* but also through all the regional papers of the British Isles, offering equipment to those who needed it. As a result *English Rainfall 1860* was succeeded by *British Rainfall 1861*, which included figures from 506 observers, and in a few more years he had 700 observers, was having to turn applicants away and had given up his job at the Board of Trade. Writing in the third person as 'the author', in the introduction to the next volume, he explained:

> Unpleasant as it was financially, he did not for a moment hesitate, rather to sacrifice his pecuniary prospects, than abandon an investigation, the ramification and importance of which we cannot yet see.

The last phrase captures much of the charm of the enterprise, viewed from a distance. Symons had no clear idea of where his efforts would lead; he just knew that it was better to have figures than not, and that the sooner it was done properly the sooner they would be useful. (Nor did he need any longer to think about Glaisher's questions, for they had been answered by the weather itself: a single wet year was enough to fill the rivers again.) Symons's force of volunteer observers could soon be numbered in the thousands, a scattered army of vicars and doctors, squires and teachers, supplemented by the occasional aristocrat, privy councillor, admiral and even two or three 'Misses'. By the time he published the twenty-fifth volume of *British Rainfall* in 1885 he felt able to claim that 'there is scarcely a spot in the British Isles from which, were I suddenly dropped from a balloon, I should not be within walking distance of one of my correspondents'.

Long before then, however, he had realized that he had to be more than a mere clearing house for numbers. The figures would be of limited use unless they were gathered according to what would now be called a standard operating procedure, and this meant enforcing discipline. Equipment, for example, had to be of the right quality. It was simple enough: usually a bottle with a funnel made in such a way that it would not admit incidental splashing, plus some reliable way of measuring the depth of what has been gathered. Readings had to be taken at the same time every day and according to the same rules, and they needed to be recorded and dispatched correctly. This would not have been possible before the nineteenth century with its railways, postal system and telegraph, but even then it was a remarkable feat. The surviving photograph of Symons, taken by one of his observers, shows a solid figure with thinning hair, grey beard, long nose and bright, cheerful eyes. The charm and ease of manner he used to get his way are evident, just as they are from his many writings, but so is the physical and intellectual effort. Whenever he could get away from the figures, this busy man visited his observers and though he was sometimes dismayed by what he found – gauges left at an angle, placed in holes in trees or propped against the walls of houses – he never lost heart. Rules were written down and distributed and those who strayed were cajoled and chivvied with great patience, while controversies over best practice were aired as fully as possible and in difficult matters even put to a vote. Symons was no dictator, and he knew that he was working with volunteers. It helped that he himself was an observer and knew the problems; his records at Camden Square had pride of place at the top of the lists every year and eventually became one of the country's longest-running series. In the same garden he also conducted many lengthy experiments with rival designs of gauge and other equipment. For some time he had a large metal tank buried to near its rim in the lawn as part of a study of evaporation.

His work had another dimension, too, for very early on he set himself the task of gathering and collating such figures as were available from before 1859. He appealed to anyone in possession of sets of rainfall observations from earlier times to send them to him and by 1865 had accumulated about 7,000 series. These were the work of curious amateurs down the ages, often fragmentary but sometimes bearing witness to a doggedness which is wonderful considering the limited use the figures can have had for their compilers. One remarkable set was maintained by Thomas Barber of Lydon in Rutland between 1736 and 1794 – almost fifty-nine years – while the earliest sustained measurements were kept by Richard Towneley of Burnley in Lancashire over most of the years between 1677 and 1705. Towneley had a clever piping arrangement from the roof of his home that enabled him to do his readings in the comfort of his own study. Of course all these series were not consistent with one another in method and they contained many gaps, but after studying them in depth Symons felt he could derive respectable rainfall figures going back to 1815. This effort has continued and today the monthly totals for England and Wales have been pushed back to 1766.

Symons supplied for British rainfall what Howard had called 'the key of analysis', and even in his own lifetime he was able to see some results. Old wives' tales and popular assumptions could be set aside as for the first time reliable generalizations about the rain were possible. The picture soon emerged of an island divided by a line that ran, approximately, from Newcastle upon Tyne in the north through Leeds, Birmingham and Oxford to Southampton halfway along the south coast. To the west of this line, including the West Country, Wales, Lancashire and, to the north, the whole of Scotland, annual rainfall was conspicuously higher than to the east. Low-lying eastern towns such as Peterborough, Ipswich and Hartlepool had less than half the rainfall of the western uplands. But there were wrinkles: wet spots in the east such as east Kent and dry spots in the west such as the Wirral in Cheshire. Manchester and Glasgow, western cities with reputations for raininess, proved in fact to be not far above national average levels. The driest place in England was the Thames estuary and the wettest by far the Lake District, where a heroic group of observers competed in all weathers for the highest readings. Like the country, the rainfall year was divided in two, with spring and summer on balance drier than autumn and winter, so that, for example, April emerged as one of the driest months – a surprising discovery for a nation which since Chaucer had associated it with showers.

Among the most remarkable things about Symons is that these great efforts were almost entirely self-financing. His father does not appear to have left the boy any legacy of substance and he gave up salaried work by 1863, after which he received a few grants from scientific bodies, which went mainly towards equipment for observers, and a couple of small private bequests. Otherwise all the money came from two sources: the subscriptions to his British Rainfall Organization, most of whose members were the amateur observers he had recruited, and the sales of a journal, *Symons's Monthly Meteorological Magazine* (fourpence a copy in 1865 or five shillings for an annual subscription). With these modest funds he ran the operation and maintained a house that was at once a library, an archive, an office, a laboratory and a home. He bought every book he could find on meteorology – the story is told of him walking into central London in the early hours to besiege booksellers before they opened in order to be sure of his purchase – and at his death his collection ran to 10,000 volumes. The reports of his observers and the historical collection were meticulously collated and maintained – they amounted to well over 100,000 yearly records and he had a special fireproof annexe built at the bottom of the garden to accommodate them. For support Symons relied at first on his mother, Georgiana, and then on his wife, Elizabeth (they had only one child, who died in infancy), but by the 1870s he was employing assistants. His thoroughness and accuracy were legendary – he took great pride in how few *errata* he had to print – while his output was astonishing. Not only did he edit both the rainfall series and the magazine, but he contributed many articles to the latter and to other learned journals while keeping up a regular correspondence in *The Times*. It can be no surprise that he was never well-off, but he complained only rarely and stubbornly rejected all suggestions that the work should be taken over by the state.

If this occurred, he wrote:

> I believe that the *esprit de corps* which at present
> exists would be extinguished, and though, by taking
> the present system and applying more red tape it
> might be possible to make the machinery even
> more nearly perfect than it is, it would be at the cost
> of that intelligent independence of thought which
> so greatly rules the progress of science.

Why did he do it? His output leaves little doubt that
Symons was one of those boy collectors who never grow
up. At the same time his tables of figures spoke to him,
offering insights large and small into the behaviour of
the British climate, and the pleasure he derived from
them shines through his writing. No, he told *Times* readers complaining of a wet summer, there had not been
more rain than usual, but more rain than usual had fallen
in the daytime, which might explain the general misery.
On another occasion he picked out the wettest day of
the preceding year and observed:

> It may be interesting to know that if the fall of
> rain on 26th September was uniform over the
> whole of London, and the area of the metropolis
> is assumed to be 78,000 acres, the water
> discharged over London must have amounted to
> nearly 3,000,000,000 gallons, the weight of which
> would be about 13,000,000 tons.

He might have added that this was about four tons for
every inhabitant. Symons was not inclined to draw grand
conclusions or state general findings on the basis of his
figures; he seems to have regarded that as premature.

Nor did he feel much need to justify his own work, although in 1880 he spoke up for his observers:

> Minor motives may have some influence, but I believe that the leading sentiment which binds together British Rainfall observers is the consciousness that they are helping gradually to store up a mass of information which is, and will yearly become, increasingly valuable to the nation at large, in relation alike to Agriculture, Sanitation and the proper appropriation of the water supply of the British Isles.

He was still working – indeed he was in his second term as President of the Royal Meteorological Society – when he died in 1900, with the fortieth edition of the rainfall series in preparation. The work carried on in Camden Square without him for nearly twenty more years before it was finally taken over by the state, and *British Rainfall* is now published annually by the Meteorological Office. Thanks in great measure to Symons, who never wanted or received so much as a penny from the taxpayer for his efforts, it is the longest continuous national rainfall record in the world.

5

When a clumsy cloud from here meets a fluffy little cloud from there, he billows towards her. She scurries away, and he scuds right up to her. She cries a little, and there you have your shower. He comforts her. They spark – that's the lightning. They kiss. Thunder!

> from *Top Hat*, the prelude to Irving Berlin's
> 'Isn't it a lovely day to be caught in the rain?'.
> In the film this scene takes place in London.

In the business of getting above the weather, and in particular above the rain, the nineteenth century saw advances to dwarf all that had gone before. By road and rail, on horseback or on foot, travel was both a drier experience and less subject to disruption. Better transport and communication meant that districts and regions were less vulnerable to the effects of weather on their crops – with the rise of trade, too much or too little rain need no longer lead to hardship or famine. At the same time advances in architecture meant that large public spaces could be roofed over, as they were in the great public halls, the railways stations and at the Crystal Palace, so that masses of people could gather and do business without risk of being soaked. The new sewerage systems, besides curtailing disease, also reduced the problem of urban flooding as the storm drains carried off rainwater. More and more, life could go on whatever the weather and, in a manner that is familiar in technology today, this process of change was accelerating. In the first years of the twentieth century the motor car became a familiar sight, and once it had acquired a roof and doors it too became a great liberator from rain. Here was a weatherproof mode of transport that could convey you, bone dry, from your own door to any other door you chose. Mass production, moreover, soon made the umbrella sufficiently cheap for almost everyone to take the risk of losing one, and by mid-century annual sales exceeded a million. The 1930s brought polythene, which had the ability to seal out rain altogether, and after that PVC and other plastics, giving us not only the plastic 'mac' but plastic windows, plastic bags, plastic soles to our shoes, plastic gutters and much else.

Science, meanwhile, was finally filling in the last gaps in the water cycle. It is hardly surprising that they were a stubborn mystery, for a great many factors were at work in the sky, some familiar and some downright perverse. Hot air rises. Moist air is lighter than dry air. The air becomes cooler as you ascend through the atmosphere. Condensation releases heat. Changes of air pressure produce cooling. Liquid water and water vapour can exist below freezing point. Many clouds never produce rain. Add to these the effects of wind and constant fluctuations of barometric pressures caused by global climatic forces and you have a picture so muddled that it seems no two clouds will ever behave in quite the same way.

The Victorians and their contemporaries solved some of the problems. For example it was a Scot, John Aitken, who in the 1880s properly explained one of the most baffling characteristics of water vapour: why is it so reluctant to condense even when it has sunk below the normal temperature for condensation, known as the 'dew point'? Aitken saw that condensation in the high atmosphere, like condensation in a steamy kitchen, needed something to stick to, but instead of a cool window or a glass containing a cold drink it had to settle for dust. The air, even the upper air, he realized, is full of motes of dust so small they are invisible, some of them salt whipped up from the sea and soil from the land, some of them pollen and some of them churned up by fires, factory chimneys and volcanoes. When the temperature falls below the dew point in a mass of air and water vapour, the vapour seeks out these motes and adheres to them. If there are not enough of them it simply waits in its gaseous state, and the temperature must fall a long way before it will agree to condense spontaneously. Water vapour, therefore, can exist below freezing point, while every tiny cloud droplet contains dust.

Other mysteries and confusions were overcome. How do clouds stay aloft? They are composed of droplets which are subject to the pull of gravity, so they *want* to fall, but they are so tiny and light they can be lifted by the slightest draught. Also, they are prone to evaporate as they descend into warmer air, the resulting vapour rising again, condensing into droplets again and descending again in yo-yo fashion, all within the same cloud. But the Victorians were not equipped to grasp all this. It was not until extensive observation at altitude became possible and until the science of physics, particularly thermodynamics, had made further progress, that a picture of what *causes* rain finally emerged. In the end there was not one cause, there were several.

What the meteorologists call a shower – a shortish burst of heavy rain – is normally the outcome of a distinct set of conditions and the process begins, like so much else, with the sun. The sun heats the earth and the air that is close to the earth is correspondingly warmed. This mass of warm air rises, taking with it a large quantity of water vapour, and as it moves up through the atmosphere it begins to cool. At a certain altitude it reaches the dew point and from here onward condensation occurs and cloud droplets form around specks of dust. These are not raindrops but a thousand times smaller, and crowded together they are visible as a *cumulus* cloud. Its base is at the dew point altitude and it may already be dark and threatening down there, while above is a churning, creamy-white cauliflower, billowing and growing all the time. Inside, droplets are bumping into each other and gaining in size and weight, but this is a slow process and it is overtaken by other events. As the air column continues to rise it passes freezing point and ice crystals begin to form direct from vapour, once again around specks of dust. The cloud can become very large, both horizontally and vertically, reaching several kilometres up into the atmosphere. It soon becomes a *cumulonimbus*, black at the foot and bulging out at the top in the thinner upper air, taking on the shape of an anvil. High up inside, water is now present in all three of its states – solid, liquid and gas. It is the ice crystals that have the advantage and they suck in much of the surrounding vapour, first from the air and then indirectly from the droplets in a process called transfer. Large crystals – snowflakes – are now forming near the top of the column and they begin to fall. As they do so they collide with each other and merge, and at lower levels in the cloud they strike water droplets, which stick to them and freeze. Then farther down, where the air is warmer, they begin to melt so that by the time they leave the bottom of the cloud they are full-sized raindrops at least one millimetre across.

Even at the height of summer, then, the raindrops in

heavy showers like this usually begin their descent as snow. *Cumulonimbus* clouds normally produce a short, intense downpour of the kind I have watched so often from my window, but they can also generate more extreme events. The electrical activity inside them may cause lightning and the rain can sometimes fall as hail, while the suction on the ground created by the up-draught of warm air inside them can, on rare occasions, pick up sand and small creatures such as fish and frogs, which later fall to earth with the rain.

A second variety of rain is known to meteorologists as *orographic*, which means that it is caused by mountains. When a mass of warm, moist air, travelling across the surface of the earth, encounters a mountain range it is pushed upwards and correspondingly chilled, prompting rain. This is not usually so violent a business as in the *cumulonimbus*; the cloud droplets tend to grow more slowly, by collision in the turbulent air, and the result is often rel-atively small raindrops or drizzle. (Technically, drizzle is formed of drops less than half a millimetre across.) These effects are extreme in the case of the Himalayas and the Rockies but even Britain's modest mountains and hills make a substantial difference. The orographic effect is strong in all the wettest places in the country – the west-ern Highlands of Scotland, the Lake District and Snowdonia – and it is one of the main reasons why the west is generally so much rainier than the east. The rela-tively low Brecon Beacons in South Wales, for example, have three times as much rain as the Thames Estuary, which lies on roughly the same latitude but to the east.

The rest of our rain is mainly delivered by frontal systems, familiar as those lines of bunting on weather maps. These are not local or even national phenomena but belong to the big picture of global weather and this, in very short order, is how they work. Along a line some-where in the middle latitudes of our hemisphere, cold air streaming south from the Pole meets warm air pushing north from the Equator and an uneasy stand-off occurs. Very little in global weather is stable, so before long kinks emerge here and there in the line and a wedge of warm air pushes northwards across it. These wedges are then squeezed, so that they become finer, like arrow-heads, with cold air on either side. At our latitude, therefore, you can have a background of cold air with an arrowhead of warm air thrusting two or three hundred kilometres northwards into it. Now visualize this whole system moving eastwards around the globe from the Atlantic to northern Europe. In, say, southern England, people are first experiencing cold, polar air, but then comes the warm air inside the arrowhead – perhaps quite a long spell of it in southern parts, with less farther north where the arrowhead narrows to its point. And after the warm air moves on towards the Netherlands or Denmark it is replaced once again by chilly polar air.

Add to this sequence vast quantities of water vapour from the Atlantic and the tropics and you have many opportunities for rain. On either side of each interface between cold and warm air, for example, there are effects likely to promote rain, while at and beyond the tip of the arrowhead there are more. Let us limit our-selves to just one of these: the moment at which the leading edge of the warm air mass arrives. This edge is known as the warm front (these phenomena were first described by Norwegian scientists during the First World War, and they borrowed the term from the mili-tary maps in the newspapers of the time). A front of this kind may be hundreds of kilometres across and the mass of air behind it may stretch back over a similar distance.

Because warm air pushes up above cold air the front arrives first at high altitude, often accompanied by *cirrus* clouds, and from this point on there is a sort of escalator overhead, with its lowest point at ground level following far behind. Warm, moist air travels up this escalator in a steady flow, and as it does so the water vapour condenses and forms into clouds. From the ground this appears to be a huge, even blanket – the *stratus* cloud – but it is not really even; it is moving up the escalator, while the whole escalator is slowly advancing. Soon light showers will begin, followed by heavier rain. The clouds become thicker, darker and lower and the rain steadily heavier, until at last the point is reached when the base of the front, or the bottom of the escalator, passes you at ground level. Then, with the polar air finally removed and the warm air in control, conditions will rapidly change, the clouds will begin to break up and chinks of blue will appear again. Sunny weather may follow.

These three kinds of rain are not entirely distinct. A good deal of our orographic rain falls from clouds belonging to frontal systems, for example, while simple wind turbulence can mimic the effects of mountains by forcing clouds upwards until raindrops form. But in broad terms these effects account for the great majority of the rain that falls on us and understanding them represents a significant scientific achievement. In particular the explanation of frontal systems, initially by those Norwegians, Vilhelm Bjerknes and Halvor Solberg, was a very important moment in meteorology, providing as it did a strong new bridge between global weather, long the subject of study, and local or national rainfall. In the years between 1800 and the middle of the twentieth century our intellectual grasp of rain, just like our physical ability to cope with it, had been transformed.

altocumulus

6

Floodwater can enter through drains, toilets and other outlets such as washing machines. The simplest way to prevent this is by putting plugs into sinks and baths and weighing them down with a sandbag or other heavy object . . . Place a sandbag in the toilet bowl and block the washing machine drain with a suitable plug (e.g. cloth or towel) to prevent backflow.

from 'Damage limitation: how to make your home flood resistant', published by the Environment Agency

The first time it occurred to me that British people might have become casual about rain was on the October day when I visited George Symons's former home in Camden Square. Apart from the plaque on the wall put there by the Royal Meteorological Society, there is not much left to see. In the garden the only relic of the great man's many instruments and experiments is a flagpole with a weathervane on top and it is looking a bit elderly now, although the owner of the house is nobly preparing to replace it with a new one. The annexe where records were stored is also still there, a solid brick structure against the end wall of the garden, but it is locked up and unused, the papers long ago removed. The house has been modernized several times since 1900 so nothing of Symons survives inside, although in the course of a recent redecoration it seems that the marks from his many bookshelves were evident in the wall plaster on the first floor. And that's it. No ghosts and no unexpected insights; Symons's legacy is elsewhere.

I was making my way home after this visit, walking towards Kentish Town with the intention of catching a bus, when it began to rain, first a drizzle and then more heavily. By way of protection I was wearing a short jacket of the kind that used to be called a windcheater. When I bought it, and when I put it on that morning, I had a notion it would protect me at least against a light rain, but it has no hood and as I look at the label now it makes no pretence of being waterproof. So I was soon feeling that chilly tingle on my shoulders as raindrops penetrated to my shirt and the skin beneath. I thought of a saying I once heard that is popular in Norway: 'There is no such thing as bad weather, only bad clothes.' By any definition I was wearing bad clothes. At that moment I could do nothing about my clothing, but I did have other options. I could, for example, have found shelter and waited, but if I considered this it was only fleetingly. I could have broken into a run. I have since read a helpful paper by two scientists which shows that this is the smart thing to do: yes, you will catch more rain on the front of your body as you run, but overall you will end up drier because you will reach shelter sooner. (The scientists proved this both mathematically and experimentally – the latter by going out in the rain together in identical tracksuits, one walking and one running, and then weighing their outfits afterwards.) At that time I had not read this paper and even if I had I would probably not have run. What I did instead was press on briskly to my destination, ten minutes away on Kentish Town Road, and it was only when I reached the bus stop that I took shelter in a doorway.

By now I was wet and grumpy, but as I waited for my bus and watched the people of Kentish Town go by I noticed something. Some of them, mostly women, had umbrellas and a few were wearing genuinely waterproof outer clothes, but a clear majority were no better prepared than I was. What was more, most of them were reacting to the rain in much the same way that I had. Men in suits simply pinched their lapels together, bowed their heads and hurried; teenagers in hooded sweat-tops wandered along oblivious to the dark patches growing on their shoulders and heads; women in coats of the non-waterproof variety held newspapers over their hair; mothers with children in plastic-covered buggies screwed up their eyes and forged on. Like me, they had undoubtedly had the same experience many times before and, again like me, they had not learned from those mistakes. Perhaps if they had been Norwegians they would have been wearing Gore-Tex from head to toe, but they were not and they just could not be bothered to take the precautions necessary to stay dry. In fact, I reflected as I watched, not being bothered about rain is the norm these days.

This is not to say that we are indifferent to it. Rain is wet; it spoils hairstyles; it leaves clothes damp and puts them out of shape; it splashes uncomfortably on your face; it spatters your spectacles; it spoils the newspaper or book you are carrying; it robs you of your composure and even, to a small degree, of your dignity. And it makes you cold, both as it strikes (it may well have been snow only moments before) and when it is evaporating off you (of which more later). All of these are reasons to dislike rain and avoid it, but they are not for the most part very big reasons and it seems that it does not take much to cause us to set them aside. Look again at those people on Kentish Town Road. Their expressions left no doubt that they would rather not be in the rain, but like me they kept on walking, finishing their journeys. Why? Well, why not? None of them can have been going far, otherwise they would have been in a car or a bus or the Underground beneath, so they knew that their exposure to the rain would be short. Certainly they would arrive at their destinations colder and wetter, but it is a safe bet that, wherever their destinations were, almost all of them would find central heating or air conditioning there which would quickly bring them back to a warm, dry state. They had decided to get on with their business, rain or no rain.

And the people on the pavement were only a fraction of all those on that road that afternoon. Many were in cars and buses, snug and dry and on their way, and many more were in the buildings on either side, some unaware of the rain and a few, no doubt, looking out and feeling glum or smug, depending on their inclination. Some of them were probably waiting for the shower to end, having deferred some errand. How much inconvenience did that shower cause in that place on that day? Very little. Extend this more widely to everyday life, at least in our towns and cities, and you see that rain is a very marginal matter. Of course it affects some professions and trades in important ways – notably builders and engineers – and of course it affects all of us in lesser ways, such as when it spoils wedding receptions or sports days or cricket matches, but taking it all together rain is normally a minor nuisance for modern Britain and no more. We flit from rainproof home to rainproof car to rainproof office and back again, stopping sometimes at a rainproof wine bar or supermarket. Once in a while we will deliberately expose ourselves to rain in the course of a healthy walk – suitably attired – and once in a while we will be caught off guard and have to rush through a shower to catch a bus or find a car parked a street away, but that is the limit of our exposure.

One reason for this comfortable state is that we no longer fear getting wet in the way that the generations who read William Buchan did, for although the link that Buchan saw between wetness and sickness may not have disappeared completely under the weight of modern medical scrutiny, it is somewhat more slender. Doubts seem to have set in with the First World War, when something odd was noticed: although soldiers at the Western Front spent a great deal of their time in wet trenches with wet clothes and wet feet, embracing all the conditions that Buchan thought most dangerous, they were no more prone to catching colds than soldiers who were behind the lines in warm, dry accommodation. If anything, in fact, the figures suggested they were less so. After the Second World War this intriguing matter was taken up at the Common Cold Research Unit, set up in a former army camp near Salisbury. This was the institution – beloved of 1960s comedians – that paid volunteers to lounge around in Nissen huts for week upon week, exposing themselves to the cold virus in a variety of conditions. Some people became addicts of the place and when turned away would come back in disguise to try again. Other volunteers fell in love with each other, sometimes behaving in breach of the unit's contact regulations. One party was even taken to a remote Scottish island and stranded there for a few weeks.

Among the experiments carried out at Salisbury was one that involved eighteen human guinea pigs, divided into three groups. Six were given a shot of cold virus up the nose and then left to their own devices. Another six 'took hot baths and thereafter stood about in bathing attire, undried, in a cool corridor for as long as they could stand it. Then they were allowed to dress but still wore wet socks for some hours.' The last group had both the virus and the ordeal by water and wet sock. The results at first seemed to suggest that more members of the last group caught colds, as might be expected, but when the tests were repeated it was the first group who caught most colds and eventually the scientists had to admit they could draw no conclusion at all. Further experiments followed, introducing a new hazard: 'The chilling consisted of sending the subjects out for a walk in the rain. They were not allowed to dry themselves for half an hour after their return and furthermore the heating in their quarters had been turned off.' Again, however, the results were unable to show any link between being wet and catching colds.

Medical science has come a long way since then, not least by transferring the punishment from statistically insignificant numbers of human beings to large armies of mice, and we are proportionately the wiser. As the Salisbury scientists knew, the common cold is a virus, the rhinovirus, and it is not 'carried' in any way by rain. Nor does rain on its own increase your susceptibility to catching it. But what rain can do is lower your temperature through the process of evaporation, because when the warmest and most energetic water molecules rise into the air off your hair, skin and clothes the average temperature of what is left behind falls quite sharply. (This is why we cool down when we sweat.) This loss of warmth reduces the efficiency of your immune system and thus increases your vulnerability to viruses in general. So there is a link, but we should keep it in perspective. Many other forces are at work in the spread of infection, not least the need for proximity with someone already carrying the virus. You run a higher risk of catching a cold, and many other viruses, from sitting with an infected person in an enclosed space, however cosy it may be, than you do from five minutes in the rain. Your age, general health, stress levels and whether you smoke or not also have parts to play.

Even when caught, colds are less worrying. Buchan described the doctor's experience in 1769:

> On examining patients we find that most of them
> impute their diseases either to violent colds
> which they had caught, or to slight ones which
> they had neglected.

Today in our centrally heated homes we are much better placed to nurse ourselves through infections and we have a great many drugs and treatments to relieve the symptoms while we do. The risk of slipping from a cold into some more serious disease is correspondingly far lower. Unless lightning happens to strike, therefore, you are unlikely to 'catch your death' by going out in a shower of rain.

A dim awareness of this may be one reason why the British are so happy to ignore that Norwegian saying about bad weather and bad clothes. We have a clear idea of what constitutes bad weather and by and large it is rainy weather – sometimes very windy or cold days qualify, but rain is the standard ingredient and rain is common. Meteorologists agree that a 'rain day' in Britain is a day when 0.2 millimetres of rain falls, and they count these days. Curiously, although total rainfall varies considerably between different parts of Britain, sometimes by a factor of four or five, the number of rain days to be expected each year wherever you may be falls within a relatively narrow range: somewhere between 150 and 220. In other words, put very crudely, most people in

Britain can expect rain around half the days of the year, with a distinct bias towards the winter months. If British people were Norwegians this fact would presumably be reflected in their dress, but they are not Norwegians.

The right clothes exist for this climate, but we don't like them. Truly waterproof fabrics, whether they 'breathe' or not, are cumbersome and cold to the touch, and by and large we avoid them because they are not worth the discomfort and the unsightliness. Those people who turn up at the office in bright, hooded outfits with zips in all directions tend to be seen as overdoing it. Those of us who wear raincoats, meanwhile, prefer them relatively comfortable, in a cloth that is at best, to use the shop assistants' jargon, 'showerproof'. They may not quite qualify as bad clothes but they aren't exactly good in rain either. Then there are hats: forty-odd years ago most men wore a hat of some kind, a practical precaution against both cold and rain, but as we know, somewhere in the 1960s men's hats went out of fashion. And they weren't superseded by something better, but by nothing at all; most men, like most women, go bareheaded into the rain. Finally, umbrellas. They are cheaper than ever and they fold up more neatly than ever, but if you watch what happens in the street on a rainy day at best one third of the people are using them.

British people, in short, are unworried by the rain. Watson-Watt said they do what they 'want to do, in spite of the weather'. I would go one short step further and say that so far as they can they ignore the weather. This is not the stereotype, of course, because they are supposed to be obsessed with it. Sooner or later in this book this idea has to be confronted, and the moment seems to have arrived.

7

Heavy rain falls at a rate of 4.0mm (0.2 inches) per hour or more; light rain at less than 0.5mm (0.02 inches) per hour and moderate rain between the two.
Heavy showers fall at a rate of 10mm (0.4 inches) per hour or more; light showers at less than 2mm (0.08 inches) per hour and moderate showers between the two.
Rain is reported as continuous when it has fallen for the preceding hour with breaks not exceeding ten minutes; it is otherwise reported as intermittent.

from *Teach Yourself: Weather*, by Ralph Hardy

It was Dr Johnson who in 1758 provided the benchmark statement, one that is quoted in almost every discussion of Englishness and of the British climate. 'When two Englishmen meet,' he wrote, 'their first talk is of the weather.' Even in Johnson's day, it seems, the question of whether it was unseasonably wet or unseasonably cold was to all appearances a national obsession, and conversation about it was part of the furniture of everyday life. And so, with reassuring constancy, it remains today. Nobody keeps count, but there can be no doubt that a remarkable proportion of encounters between British people must still begin with some exchange about sun, rain, cold or cloud before the conversations depart in other directions. We all know, too, that the British are devoted to their weather forecasts, so by common consent this weather fixation is a cosy national quirk, one of the things that makes British people British.

Johnson's remark comes from the opening of an article in *The Idler* magazine in which he attempts to justify and explain this characteristic of his countrymen. It goes on:

> An Englishman's notice of the weather is the natural consequence of changeable skies and uncertain seasons. In many parts of the world wet weather and dry are regularly expected at certain periods but in our island every man goes to sleep unable to guess whether he shall behold in the morning a bright or cloudy atmosphere, whether his rest shall be lulled by a shower or broken by a tempest. We therefore rejoice mutually at good weather, as at an escape from something that we feared, and mutually complain of bad, as of the loss of something that we hoped.

The climate is of concern because it is so variable, he is saying, and this is an idea that seems logical. It makes a pretty picture too, all those Englishmen greeting each other every morning with mutual congratulations on the absence of rain or mutual laments at the absence of sun. But it doesn't ring true. Certainly today people don't feel relief or disappointment in that way, and yet they still talk about the weather, and even applied to the eighteenth century it is hard to believe. The more so because Dr Johnson didn't believe it himself, for we know that in practice he was nowhere near so reasonable about this business as his published words suggest. His friend Dr Burney recorded that if anybody mentioned the weather in Johnson's hearing the great man would fly into a rage, snapping: 'Poh! Poh! You are telling us that of which none but men in a mine or a dungeon can be ignorant.' So far as he was concerned it was not a fit topic for grown men to discuss.

Two centuries after Johnson, the Hungarian George Mikes picked up the subject in *How to be an Alien*, a book ostensibly addressed to his fellow immigrants. 'You must be good at discussing the weather,' he wrote, for 'in England this is an ever-interesting, even thrilling topic.' To help his students he supplied some model dialogues to be learned by heart, of which this is one for wet conditions:

'Nasty day, isn't it?'
'Isn't it dreadful?'
'The rain . . . I hate rain . . .'
'I don't like it at all, do you?'
'Fancy such a day in July! Rain in the morning, then a bit of sunshine, then rain, rain, rain all day long.'
'I remember exactly the same July day in 1936.'
'Yes, I remember too.'
'Or was it in 1928?'
'Yes, it was.'

Mikes advised his readers that if they learned no other conversations but these and repeated them at every opportunity, 'you still have a fair chance of passing as a remarkably witty man of sharp intellect, keen observation and extremely pleasant manners'. But the really vital thing, he stressed, was that 'you must never contradict anybody when discussing the weather'. He was surely right, or at least more right than Johnson: there is no point in having a view because these conversations are not really exchanges of information and nor do they express real opinions or emotions; they serve a different purpose. In *The Importance of Being Earnest*, when Jack remarks to Gwendolen on a charming day he receives the rebuke:

> Pray don't talk to me about the weather, Mr
> Worthing. Whenever people talk to me about the
> weather I always feel quite certain that they
> mean something else. And that makes me so
> nervous.

Jack, of course, did mean something else – he meant to propose marriage – but he wasn't quite ready for that and the weather sprang to his mind as a stopgap. It did so because, since it is at once so changeable and so moderate (Johnson was right about that), there is always something to be said about it, however uninteresting, and yet there is very little danger that what you say will excite controversy. That, surely, is the true role of the weather in British conversation: a sort of default mode enabling people to talk when silence is impossible and diffidence or manners block the way to a more meaningful exchange. In waiting rooms and on buses, in contacts with half-recognized acquaintances and with the shy or preoccupied, it is acceptable and even necessary to say: 'Nasty day, isn't it?' In short, while British people do talk about the weather that does not mean they care about it or are interested in it.

Nor do high viewing figures for television weather forecasts prove a preoccupation with the subject. My own experience, at least, suggests something different: if, two minutes after the forecast was over, you were to ask me what the forecaster had said I would usually have no idea. My eyes may have been on the screen but I was not listening, or if I was listening I was not digesting the

information. As for the images, they may be designed for
idiots these days but unless there is something extraordinary going on they still do not register. I have tested
this on friends and found that even the most attentive of
them took in little more than I did, usually losing interest once the forecaster stated whether or not it was likely
to rain tomorrow. Of course this is a city or town
dweller's view, but then that's what most of us are, and it
falls some way short of an informed interest in the subject. In any case, even my limited experience of
foreigners and their weather forecasts suggests that there
is nothing special about British interest levels. For example, where BBC and ITV place the forecasts *after* the
main news, French television stations often include
them at the top of the bulletin. Seconds into the programme and alongside the headlines, therefore, the
announcer is telling you what the weather will be like
tomorrow. As for the Americans, they have entire chan-
nels devoted to the weather, something that has yet to
happen here.

If the British are not really obsessed by the weather,
does the weather influence them in other, less obvious
ways? Does it help shape the national character? When
Jeremy Paxman was writing a book on the English a few
years ago he was naturally curious about this, but he
found that all the readily available evidence was contra-
dictory. For example, the winters are thought to be
depressing, yet it is in the spring and early summer that
suicides reach their peak. Similarly, while some have
suggested that the English climate was bracing and
invigorating, eighteenth-century French visitors were
struck by its energy-sapping melancholy. Paxman's
investigation did not get far, for on consulting an academic he was told that while the question might make

an interesting research topic it would be frowned on by the scientific establishment and 'there would be difficulty funding it'.

This is a point worth dwelling on. There is good reason for this fastidiousness, since studies of the relationship between climate and character have been badly tainted by association with racism and Nazism. Most prominent in this field in the first half of the twentieth century was an American, Ellsworth Huntington. A geographer and explorer who travelled widely in central Asia as a young man, Huntington later settled at Yale University, where he taught for many years. His earliest travels pointed him towards the ideas that dominated his life. In his first book, *The Pulse of Asia*, published in 1907, he wrote: 'Today the strongest nations of the world live where the climatic conditions are most propitious.' In a dozen subsequent books with titles such as *Climate and Civilization* and *The Character of Races* he pursued the notion that nations and races were shaped in large measure by their climates. To prove his point he measured the performance of factory workers and students in varying conditions of temperature and humidity. For mental activity, he found, 'the ideal condition seemed to be mild winters with some frosts, mild summers with the temperature rarely above 75 degrees, and a constant succession of mild storms and moderate changes of weather from day to day'. Transferring this to a map of the world, he showed that these were precisely the conditions of northern Europe and the north-eastern United States, in his view self-evidently the most productive and dynamic societies in existence.

Appealing as they did to the vanity and prejudice of his natural readership, Huntington's books sold well and his ideas travelled widely. His scientific credibility, however, was always in doubt since his techniques were often comically inadequate. He once conducted an international survey of intellectuals and was disappointed to receive few replies from thinkers in what we would call the developing world. Putting this down to their inability to understand his six-page letter in English, he blithely proceeded to make his calculations about the global distribution of intellectual capacity without their assistance. In the same vein, when he constructed a league table of the places whose climates were most conducive to inventive thought it was doubtless no surprise to him that Connecticut, the home of his own university, should have emerged as the closest to perfection.

After the Second World War this whole field of 'climatic determinism' was discredited and it became a pariah topic; to most scientists the underlying assumptions of any possible research were too abhorrent to be entertained. Little wonder that Paxman's academic was doubtful about funding. And yet the subject is not entirely closed. Modern historians, for example, while they have no time for any notion of the predominance of climate, tend to accept that climate is one of many factors influencing the course of history and thus – as Huntington might have put it – of nations. People in different climates, after all, are likely to have some different priorities and those priorities may influence what they do.

Where does this leave the British and their rain? Frustrated, I suspect, as Paxman was. Often in the course of my research people would express curiosity about the influence of rain on the national character, although few of them could suggest a line of cause and effect. It seems to be a natural subject for speculation. Some people

mentioned Seasonal Affective Disorder (SAD), the afflic-
tion that leaves many people depressed during the dark
winter months, and it is certainly tempting to imagine
that Britain's heavy winter cloud cover, and consequent
rain, play a part in darkening the national mood. The scientific evidence, however, suggests that this is probably
not the case. The condition was discovered in the United
States in the mid-1980s when it was found that the proportion of sufferers from certain depression-linked
symptoms rose steadily with the latitude – in other words
there were relatively few in Florida, and progressively
more as you moved north towards the Canadian border.
That strong connection with latitude remains the key to
the condition, suggesting that the most important factor
is the number of hours of daylight experienced by sufferers, rather than the quality or intensity of that daylight,
which might be affected by clouds. So clouds and rain
seem to play little part in SAD.

There may well be something in this question of
weather and character, but it must be elusive and if it
exists it is likely to be complex and subtle. British rainfall, after all, varies quite strongly with geography so that,
for example, any condition or characteristic it aroused in
the English would presumably be noticeably more
intense in the Scots and Welsh. By the same token, identical conditions or characteristics would surely be present
in people in other parts of the globe who experience the
same sort of rainfall – or maybe not, since so many other
forces could be at work. In the end the scientists are
probably right: it's so fluid a subject it's hardly worth
investigating, and somehow the more you talk about it
the dafter it sounds – you risk slipping into the strange
world of Ellsworth Huntington. So let's leave it there.

cumulus humilis

8

Be still, sad heart, and cease repining;
Behind the clouds is the sun still shining;
Thy fate is the common fate of all,
Into each life some rain must fall,
Some days must be dark and dreary.

from 'The Rainy Day', by
Henry Wadsworth Longfellow

The first practical steps towards scientific weather forecasting were related directly to advances in
communication. Once it became possible for messages
to travel faster than weather, which is to say once the
telegraph came into service, it was a simple matter to
warn the people next in line what was happening over
your head. The desire to do better than this was strong,
particularly as the increasing understanding of the
weather in the nineteenth century constantly hinted that
it followed some logic, however vast and complex that
logic might be. By the time of the Great Exhibition in
1851 the first maps were appearing showing the distribution of barometic pressure – the lows and highs – and
these gave an inkling of the prospects for wind and rain.
Over the next century meteorologists built on this foun-
dation. Observation points and sources were distributed
more and more widely, especially to the west in the
Atlantic, and the data they provided was mapped and
recorded so that patterns of weather behaviour could be
established. When frontal systems were explained in

1917 this was an enormous boost – anything that is systematic, after all, must be predictable to a degree. In time a good forecaster was able to look at the current map and its evolution, apply his knowledge of similar weather states in the past, adjust to take account of existing variables and produce an intelligent prediction for tomorrow's weather. As with all forecasting, the farther ahead he tried to think the less reliable his thoughts became, and he would be reluctant to offer opinions much beyond forty-eight hours ahead. Such forecasts were undeniably subjective and fairly vague, but they became sufficiently valuable to play an important part in the Second World War, the high point of their achievement coming with the successful identification of a window of calm weather for the D-Day landings.

Long before then, however, meteorologists had seen that something better might be possible, and here we meet another remarkable Englishman, Lewis Fry Richardson. Born in 1881 into a Quaker family in Newcastle upon Tyne, where his forebears had been tanners for three centuries, the young Richardson managed to find his way to Cambridge University, where he developed interests in meteorology and mathematics. These he pursued at the Meteorological Office, which posted him in time to what he called the 'bleak and humid solitude' of an observatory at Eskdalemuir in south-western Scotland. A pacifist, he joined the Friends' Ambulance Unit in the First World War and served as a driver on the Western Front, where, almost incredibly, he devoted his spare time to a monumental effort of calculation which he hoped would demonstrate a better way of forecasting the weather. The existing techniques relying on maps were, as he put it, 'based upon the supposition that what the atmosphere did then, it will do again now', but he knew that if experience showed anything it was that nature rarely if ever repeated itself. If weather forecasts were to improve, therefore, if they were to acquire greater accuracy and greater range in time, they would have to operate on different principles, and what Richardson suggested was that mathematics held the key.

Weather, like so much else, is ruled by the laws of physics and the working of these laws can be represented mathematically. It follows that if you have enough data of the right kind you can describe in mathematical terms what is happening in the atmosphere, and since these atmospheric events are in effect scientific processes – movements of air and vapour, changes of temperature and so forth – you should be able to predict, with some accuracy, what will happen next. Of course there was no possibility of Richardson having enough data of the right kind at that time, even if he had not been sitting in a war zone, and therein lies much of the charm of his work. He was attempting to prove the possibility of something for which science and technology were simply not ready, in the vague hope that perhaps one day they might be. Using a set of weather data for central Europe on a day for which unusually full figures were available – 20 May 1910 – he calculated what, applying the laws of physics, should have happened next. Given the number of variables and the scale of the subject this was an exercise of great complexity, and when he completed the task his outcome, as it happened, was completely wrong: it bore almost no resemblance to how the weather had unfolded on that May day. But Richardson was clever enough to see that this made little difference, that his idea was a good one and his principles were broadly sound, so he published it all in 1922 in a book entitled *Weather Prediction by Numerical Process*.

Cheerfully, he reflected on the scale of the effort involved in proper forecasting by his method:

> It took me the best part of six weeks to draw up the computing forms and to work out the new distribution in two vertical columns for the first time. My office was a heap of hay in a cold rest billet. With practice the work of an average computer might go perhaps ten times faster. If the time-step were three hours, then thirty-two individuals could just compute two points so as to keep pace with the weather.

By 'computer' he meant a person and not a machine, and by 'two points' he meant the development of weather between two points on the map. Since the objective was to compute the weather for the whole globe this effort would have to be multiplied, by his estimate, 2,000-fold; in other words it would take the combined efforts of 64,000 people *merely to keep up with the weather*. In fact it has since been established that this was another thing that Richardson got wrong, and that, accepting his reckoning of the work rate per individual, it would actually have taken 204,800 people to do the job. Richardson thought even his own number was 'staggering' but all that was needed for such enormous calculations, as we now know, was the electronic computer, and thirty years later the first of these became available. Today all the world's serious weather forecasting depends on electronic computers employing Lewis Richardson's numerical principles (if not his actual equations). The Meteorological Office at Bracknell, for example, prides itself on its two Cray-T3E supercomputers, the most powerful machines of their kind in the world.

As a result of this scientific and technological leap, coupled with powerful observational techniques such as radar, weather forecasting has improved beyond recognition since the 1960s and it is now an invaluable and generally reliable tool in many branches of human activity from air travel to fishing to fighting wars. Today's three-day forecast is more likely to be correct than a one-day forecast of forty years ago, while today's one-day forecasts are far more geographically precise than could have been dreamed of then. With due credit to Richardson this represents a triumph, a milestone in getting above the weather. But the popular perception of forecasting is different, and for interesting reasons, I believe.

Of course there are still mistakes, occasionally spectacular ones, but they are rarer and rarer. The real problem with the forecasts is to do with scale and it reflects the subjective character of rain. Ideally, what would we like to know about tomorrow's weather? First, perhaps, will it be warm or cold? This much we can be told with some accuracy. Then we want to know, not just whether it will rain, but whether it will rain *on us*. To be of most practical assistance to ordinary city and town dwellers, a forecast should be able, reliably, to allow them to plan not just their clothing but their movements. Will it be raining downtown at 9 a.m., the time when I normally leave the car park and walk to the office? Will it be raining there at lunchtime, when I want to pop out to the supermarket? Will it be raining in the park on Saturday morning, when I normally go for a run? This may seem to ask an absurd amount of the forecasters, but if the answers were available we might well adjust our plans accordingly. They are not available and not likely to be for some time, even though meteorologists are beginning to speak of 'postcode forecasting'. It is just too difficult, too individual. On the one hand most of us pass through several postcodes in a day and often change our plans as we move, so our personal forecasts would be complex, and on the other it is doubtful if they can ever be that precise – when I visited the Met Office I was awestruck by the sophistication of their methods but I still came away with a quote in my notebook saying: 'Of course computers are not very good on the actual progression of rain.'

The real point, however, is that we don't expect this kind of service and we are in no special hurry to have it. Rain – see Kentish Town Road – doesn't bother us that much. If it did, for one thing, we would know more about Luke Howard's cloud classifications and about the different kinds of rain, because that alone would make us, without any help from supercomputers, reasonably competent forecasters over periods of a few hours. Somewhere at the backs of our minds, I suspect, we take comfort in the thought that forecasting is now an advanced science and that mankind has an ever-increasing ability to anticipate the weather, but as a practical tool in everyday life the ability to predict rain is not important to us. If a shower comes along then we will avoid it if we can, and if we can't we will endure it and dry off as soon as possible. It is only when things go wrong that our dim consciousness of scientific meteorology rises to the surface, so that an unannounced gale or storm becomes as much the fault of the forecaster as of the weather. 'All those bloody computers and radar pictures,' we grumble, 'and they can still miss that!'

9

Why does it always rain on me?
Even when the sun is shining,
I can't avoid the lightning.
Oh, where did the blue skies go?
And why is it raining so?

'Why does it always rain on me?' was a hit for
Travis in 1999. When they played it at the
Glastonbury festival the clouds opened
and it rained on everyone.

The village of Farringdon stands on the A32 road in
Hampshire, about forty-five minutes' drive south-west
from the outskirts of London. A little to the north is
Chawton, once the home of Jane Austen, while to the
south lies Selborne, where Gilbert White kept his diary,
so we are in a sort of English literary heartland. White
and Austen might still recognize the green farmland
and woods that roll away in all directions, but as a
recent history of Farringdon observes, 'the last century
witnessed more changes than the previous 700 years'.
In familiar terms it goes on to lament this modern
transformation:

Employment opportunities within our villages
have declined. Now we mostly have to find
employment in nearby towns or cities. There is a
mass exodus of workers all having to travel by car

to work or catch the train to London. Family time is restricted to weekends.

The village has no shop and no bus service.

In fact there are two Farringdons, half a mile apart. Upper Farringdon is a pretty collection of comfortable houses and cottages clustered around a medieval church and an extraordinary red-brick town hall, while Lower Farringdon, a ribbon development along the main road, is less picturesque and more workmanlike. It is Lower Farringdon, in the valley, that concerns us. The contours on the Ordnance Survey map tell the story: the 120-metre line describes a long, narrow loop between hills that rise on either side to 160 metres and more. The A32 road runs along the bottom of this gully, rising to cross the line just to the south of the village. You might expect to find a trace of blue running along beside the road, but there is none. Lower Farringdon, according to the map, has no stream and no river.

It was raining very heavily as I drove into the village from the north one Monday morning, and the gutters were awash. I passed the pub, the sign for Farringdon Business Park and the imposing corner building which was obviously once the shop, and I pulled in on the left. There, straight in front of me, was what I had come to see: a cluster of modern houses. Eight years before, with the approval of a local council that was eager to see new, low-cost accommodation in the area, a company of developers built sixteen semi-detached houses here. The site was, ostensibly, a prime one, in the middle of the village, tucked into the angle between the A32 and the side road that leads up the hill to Upper Farringdon. Now as I looked at Chase Field, as the development is called, the two entrances were closed by high fences and the scene was deserted. All the ground floor windows of the new houses were blanked out with chipboard and the one car occupying the generous, brick-paved parking space in front was a red Mini darkly stained with green slime and obviously abandoned. Chase Field was comprehensively flooded in December 2000, just a week before Christmas, and had lain empty ever since.

Clive Butcher and his wife Donna, who lived in one of the flooded houses, told me what happened. On the Friday the water had risen close to the level of the front doorstep but they didn't really believe it would go any higher and they remember joking with neighbours about it. Sensibly, though, they spent the day and the evening moving their belongings upstairs and sandbagging the front and back doors. The sofa and armchairs they raised up on bricks and the washing machine they lifted on to the kitchen work surface – 'I think it's the adrenaline,' said Donna, 'you do things you wouldn't think you could.' By midnight, however, the battle was lost. 'It just invaded,' said Donna, still half in disbelief. Clive explained: 'It didn't come through the door. It came up through the holes in the floor for the radiator pipes. By midnight we were sloshing around up to our ankles.' Donna went to bed upstairs, with the children and all the movables, while Clive struggled all night to limit the damage below. There wasn't much he could do to make a difference, for before long the water was almost waist deep inside the house – high enough to soak the sofa on its plinth of bricks – and on the Saturday morning they and all the other residents of the little close were evacuated in rubber dinghies. Their Christmas was ruined and they faced months of disruption and endless miserable haggling with insurance companies, the council and the developers. The Butchers were eventually rehoused in one of four of the new houses that face the road to Upper Farringdon; these stand on slightly higher ground and narrowly escaped flooding. Other former residents were scattered around the countryside in what accommodation could be found.

What went wrong in Chase Field seems pretty clear, particularly if you talk to Tim Cherrington, a Farringdon parish councillor, local historian and resident of sixty-three years' standing. The Ordnance Survey map may not show a river in the valley, he explains, but there is one; it's just that there is usually no water in it. When there has been a prolonged rainy period, however, it fills up and flows northwards towards Chawton. It even has a name: the Lavant Stream. And to prove to me that this was not just folklore, Mr Cherrington produced a wad of old photographs and graphs. One of the photographs, an aerial shot, showed the Lavant Stream weaving its marshy way through fields in the direction of Chawton. It certainly looked like a river. Another showed the site of the new houses – before they were built – as a pond with a pair of swans on it. It is just about the lowest point in the village, and in fact gravel was dug there at the beginning of the last century for use in building the now-defunct Meon Valley Railway, so the site is even lower than nature made it. Every so often – perhaps every five or seven years – there would be a wet winter and in February or March that dip by the crossroads would fill up. What was happening was not flash floods, with water running off the hills and filling the gully below, but a rise in the groundwater level, the point beneath the surface at which the earth is saturated.

Mr Cherrington's graphs were records of groundwater levels going back to the beginning of the 1990s. The line showed a vigorous zigzag with an annual low point of around ninety-five metres above sea level, usually reached in the autumn. Winters being wetter, the peak was normally touched in March, when a level of a little more than 110 metres was usual. Remember now that looping 120-metre line on the Ordnance Survey map; the very bottom of the valley as it passes through the village must be below that point. Sure enough, flooding begins when the groundwater reaches 116 metres. At that point ditches and low-lying gardens fill up, as does the lowest point of the Chase Field dip, which is the little roundabout in the middle of the cluster of houses. Back in 1995, two years after the houses were built, that point was passed and the water rose almost to the doors. When residents expressed alarm the developers said it reflected 'exceptional weather conditions' which were unlikely to be repeated for a very long time.

What happened in December 2000 was much more extreme than 1995. The groundwater level reached 118 metres, or hip-height in the houses, and remained within a few centimetres of that *until late April*. So not only did it rise higher than on any occasion in recent memory, but it did so at least three months earlier in the season and remained that way for four months. Again, not unreasonably perhaps, it was described as a freak, but when I visited Lower Farringdon in mid-October of 2001 the groundwater level was once again relatively high for the time of year, at ninety-nine metres. When I talked to the Butchers and their neighbour Louise Goddard-Jones it had been raining heavily on and off for three days and they were worried. 'Don't think I'm stupid,' said Ms Goddard-Jones, 'but it's really *wet* rain. Just step out in it and it soaks you through.' Donna Butcher agreed: 'It's not ordinary. It's like something from the tropics. It falls out of the sky like sheets.' I had to agree too. I had been walking around in it, looking at the village, and though I was wearing my bright blue, hooded waterproof jacket, my trousers were wet through below the knee. But at least I could drive away. Rain for these people was not just a matter of damp clothes and they worried about it a great deal. They knew the groundwater level; they knew it was still only the autumn and they could see the rain falling. What did I think, they asked. Would there be more floods? I could not say. When I got back to my car it was in a huge puddle, while out in the middle of Chase Field, behind the barriers, another great pool of water had formed.

I have suggested above that most of us are casual about the rain these days; obviously that is not true of the Butchers and their neighbours or of the thousands of other people around Britain who are vulnerable to groundwater flooding. But Lower Farringdon provides an illustration of another kind of casualness, even a complacency. There is no dispute that the houses were built at the lowest point of the village, where in the past there had been flooding. There is no dispute either that some in the village advised against it. New houses were needed, however, and the building went ahead in the confidence that floods severe enough to enter the houses were rare and that modern drains would normally be up to the job. After the near-miss in 1995 the drainage in Chase Field was supplemented with an emergency pump and holding tanks, but in December 2000 these were soon swamped. The power of the rain and the rising groundwater, in short, had been badly underestimated and the power of civil engineering overestimated.

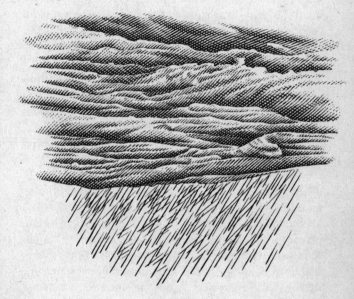

nimbostratus

10

I do not know much about gods; but I think that the river
Is a strong brown god – sullen, untamed and intractable,
Patient to some degree, at first recognised as a frontier;
Useful, untrustworthy, as a conveyor of commerce;
Then only a problem confronting the builder of bridges.
The problem once solved, the brown god is almost forgotten
By the dwellers in cities – ever, however, implacable,
Keeping his seasons and rages, destroyer, reminder
Of what men choose to forget. Unhonoured, unpropitiated
By worshippers of the machine, but waiting, watching and
  waiting.

from 'The Dry Salvages', in
T. S. Eliot's *Four Quartets*

Lewes (population: 15,700) is an attractive Sussex town
with a Norman castle and a rich history that includes a
battle in 1264 involving Simon de Montfort, as well as
strong associations with John Harvard (of university
fame) and that great troublemaker Tom Paine. It stands
about halfway along the hilly range of the South Downs,
astride one of several English rivers called the Ouse.
This Ouse drains a catchment of prosperous farmland to
the south of London and then winds across a broad
floodplain towards the outskirts of Lewes, where the
plain narrows suddenly to squeeze through the South
Downs. Once past the town it again meets open floodplain which takes it away to its final destination, the
English Channel. Lewes, in other words, stands at a

natural bottleneck and so it is no surprise that the town has a history of flooding. The records show that it has been flooded on at least forty-four occasions, of which the worst appear to have been 1671, 1726, 1772, 1801, 1814, 1852, 1865, 1911 and 1960.

That flood in November 1960 was an especially cruel one in that it came at a time when extensive work was being done to improve flood defences along the Ouse, but unfortunately these had not quite reached Lewes. What followed, however, was an intensified effort to ensure that an event of similar magnitude – it was said to be a one-in-100-years occurrence – would leave the town untouched. These impressive works had the additional benefit of releasing development land behind the protective barrier at a time when such land was desperately needed. Lewes was growing and expansion outwards and upwards on to the Downs presented many difficulties, so the old floodplain near the town centre soon filled up with industrial units, shops, offices and public services as well as some new housing. Since the fear of flood had not entirely gone, some of these properties were built on artificially raised ground.

September 2000 was a rainy month across England and Wales, and October began in the same wet mood, so that after a week the Ouse catchment area was fairly soggy. On 9 October the first of a series of cold fronts swept over Britain, bringing a sequence of heavy downpours to the area – in ten hours one third of the average rainfall for the whole month fell – and the Environment Agency issued its first flood alerts. The following day brought more of the same – about a quarter of the average monthly rainfall – and the first sandbags were distributed in parts of the upper catchment. The 11th brought a break in the rain and the river levels slowly fell, but as evening approached a very unusual meteorological event occurred. Two large areas of low pressure over central England and Scotland converged and came to a halt, creating in effect a very powerful and stationary force of suction. Drawn towards this from the south was a mass of warm, moist air from the Bay of Biscay, which crossed the Channel, met the rising ground of the South Downs and was pushed upwards into the higher atmosphere, where it collided with a mass of exceptionally cold polar air. The result was great turbulence and rain, and the rain was even heavier than what had gone before. In sixteen hours overnight almost nine centimetres (three and a half inches) fell, twice the total for the 9th and 10th and about the average for the whole month. With this the flood alert became an emergency.

Upstream from Lewes only one town in the catchment was vulnerable, Uckfield, and by 10 a.m. water almost two metres (six feet) deep was rushing through Uckfield's streets with such force that powerful rescue dinghies could make no headway against it. Although this caused considerable damage the peak passed quickly and the waters soon receded, so that the work of clearing up was able to start by late afternoon. The surge of water, however, continued southwards, met similar surges from other tributaries of the Ouse and then gushed out on to the floodplain above Lewes. A five-kilometre-long (three-mile) stretch of flat farmland swiftly filled up with swirling brown water to become a lake dotted with trees and brimming with filth and rubbish swept off the catchment – including a lot of stock from an Uckfield supermarket. All morning the water level rose, and as it did so the pressure built up around Lewes, testing the defences built in the town after 1960. There was still hope that they would be up to the job.

The Ouse is tidal at this point and the tide was coming in; when it turned at around 12.30 p.m., it was suggested, levels would fall. The flood, however, was beginning to find other ways of penetrating the town. Storm drains and sewers 'backed up', pushing water into low-lying streets where puddles became ponds and ponds grew into shallow lakes. Water seeped through air bricks into houses, cellars filled up with groundwater and domestic lavatories brimmed over. At 11.30 a.m., concerned that a defensive wall near the town centre might fail, the police decided to evacuate the central area that was most at risk. As 12.30 came and went, still the river rose, while people in threatened parts of the town made what preparations they could. Then, between 12.45 and 1 p.m., water began to lap over the defences, walls suddenly turned into weirs and from that moment things moved very rapidly.

Many people in different areas later described a 'torrent' or a 'wave' of water passing down their streets, and in fifteen minutes or half an hour most of the low-lying districts, from the eighteenth-century shops of Cliffe High Street to the new business parks, were a metre or more deep. At this point, due to 'significant personal risk', as the official report put it, 'the police abandoned the centre of the town and the evacuation became a rescue operation'. The hoped-for relief from the turn of the tide never came (in fact the flood was so powerful it overwhelmed all tidal effects for several days) and the water kept building up at the Lewes bottleneck. Besides shops and homes, many landmark buildings and important services were swamped. Harvey's brewery was awash, as was the new Riverside surgery and the huge nonconformist Jireh chapel, where the timber ground floor, complete with pews and pulpit, was lifted clean off its base. Tesco, Iceland and Safeway supermarkets were flooded, along with the town's bus station, the railway station, the ambulance service headquarters, the magistrate's court, a petrol station, the sewage works and the pumping station. The fire service headquarters, including a modern communications centre, workshops and stores, was a metre deep in water, while three fire engines and eight light vehicles were wrecked. The water, by all accounts, was disgusting: a cocktail of mud and flotsam with oil and diesel fuel and animal and human waste.

For eight hours the waters rose, eventually reaching a peak that was almost a metre higher than the 1960 level. It was not until after 9.30 p.m. that evening that it began to subside, and then a new problem appeared. Having passed over the defences largely without damaging them, the water was now trapped behind them and could not escape to the second floodplain below the town, so it remained knee-high and worse for the whole of that night, the next day and the next night. Very slowly it seeped or was pumped away, and it was not until the afternoon of the 14th, more than forty-eight hours after the flood began, that people were able to return to their properties, finding them stinking, sodden, filthy and for the most part, at least on the ground floor, ruined. A total of 613 homes, 207 business premises and seven public buildings had been flooded, while 503 vehicles were written off.

The official report speaks of 'a sense of grief, almost bereavement' among the victims in the weeks and months that followed, and this was deepened by a continuing fear through that winter and spring that the flood waters might return. This and the inevitable struggles with builders, public officials and insurance companies 'made it difficult always to maintain a positive attitude to recovery'. A full year later there were homes still uninhabitable and shops still shuttered. Some businesses closed for good, their owners simply giving up, and there were marital break-ups and suicide attempts. Besides the despair, the report says, there was also

> an intense anger and bitterness amongst many at the perceived failings of the 'powers that be' variously to: adequately maintain the watercourses; prevent development on the floodplain; provide adequate flood warnings; provide post-flood support and assistance; 'come clean' about the causes of the flood; or, do something to prevent future flooding.

A strong suspicion took root that the devastating waves seen by so many had in fact been a single wave, caused because a sluice gate on the upper Ouse had been opened, either through incompetence or deliberately to limit the damage to Uckfield.

Let us take a step back. This was a miserable, even a horrible event and the people of Lewes and Uckfield deserve our sympathy, but in most respects it was not unusual. Floods happen somewhere in Britain most winters and Lewes had had the experience before. No one was killed or seriously injured and in many ways the town has made an impressive job of recovery. Even the cost, which was calculated for the whole catchment, including Uckfield, to have been about £1bn, was not so very high. This was not Mozambique or Bangladesh; if it was a disaster it was on a domestic, British scale. But it *was* significant, and what made it so was what it revealed about human attitudes, and in particular scientific and official attitudes. These emerged very clearly in that official report, compiled for the Environment Agency by an engineering firm, Binnie, Black and Veatch. The authors took care to record the feelings of the flood victims, as summarized above, but their own conclusions about causes and meanings were very different:

> Having studied the event in some detail, we believe that the 12th October 2000 flood in the River Ouse catchment was the result of an extreme natural event, which was neither caused by, nor could have been prevented by any individual, organisation or action. We believe that the devastating *effect* of the flood was principally a function of: a) the characteristics and severity of this natural event; b) the natural characteristics of the catchment itself; c) the widespread development, over many years, of the natural floodplain.

In other words there was no human failure in the short term and no individual or group of individuals could be blamed. The Meteorological Office, the Environment Agency, the local authorities and the emergency services did all they could in the circumstances and the story of the opening of a sluice gate was exposed as a myth – all the relevant gates had been opened days before 12 October. But although there was no negligence or incompetence of that kind the report still pointed to a disastrous human intervention:

> For centuries the only building on the floodplain was related either to specific river-based industry such as mills and barging inns, or at commercially

attractive river crossing points. Periodic flooding would therefore have had comparatively little impact, and only the most extreme events would probably have been considered noteworthy. It was not until the economic and population expansions that came with the Industrial Revolution in the 19th century that major land drainage and river engineering works were undertaken which enabled widespread building on the floodplains. It is interesting to note that it is about this time that there start being regular reports of flooding in Lewes. More recent development on the floodplain, beginning in the 1960s, has had a particularly significant effect on the impact of extreme flood events.

This could be vividly demonstrated. About 830 properties were badly damaged by the Lewes flood of 2000. Taking into account the town's history and the lie of the land, it could be shown that if the flood of 1960 had been of exactly the same magnitude about 550 properties would have been damaged, and if the same flood had occurred in the middle of the nineteenth century the total would have been 200. And it was not just the number of buildings on the floodplain that was the problem, but the way in which they were built. Their 'extent, location and orientation' had the effect of increasing water run-off, which meant the flood reached a higher peak and lasted longer. In particular, of course, the scale and strength of flood protection added to the difficulties by preventing the waters escaping rapidly, as they did in Uckfield. Even the new buildings built on raised foundations made their contribution to the damage by interrupting flow and trapping water behind them.

When Lewes was flooded in 1960 the response was, in effect, to blame the rain and try harder to tame the waters. This time the answer was very different: we have got this wrong and made this flood worse for ourselves, the report said; we have followed a logic that caused more misery and damage rather than less. Not only should we not have built on the floodplain, but the very act of confining and enclosing the river meant that, when an extreme event came along, the consequences were even more violent. This is a striking change of attitude; what prompted it?

11

*Surveys of cloud cover in European representational-style
paintings of various periods have shown averages of nearly 80
per cent cover in pictures from the period 1550–1700, 50–75
per cent at various times in the eighteenth century, 70–75 per
cent in Constable's and Turner's time (1790–1840) and 55–70
per cent in the twentieth century.*

from *Weather, Climate and Human Affairs*,
by H. H. Lamb

Back in 1859 when George James Symons began his collection of rainfall figures, he was responding to public
fear of a lasting change in the climate. There had been a
run of dry years and, in the absence of adequate statis-
tics, no one could tell whether the drought had come to
stay. Within a few decades his work provided scientists
with a new, much fuller perspective on British rain and
what they saw reassured them. Yes, there were occasional sustained spells of rain and drought just as there
were of heat and cold, but it seemed that everything in
the British climate levelled out in the long run.
Centuries of anxiety – the 1859 scare was one of many –
were now put down to ignorance, to the difficulties of an
age when people lacked the 'key of analysis' provided
by Howard, Symons and their kind. The climate was
stable. This notion proved not only reassuring but con-
venient. Civil engineers building roads, bridges,
reservoirs, railway embankments, sewers or any other
large structures needed, as we have seen, to calculate

average and maximum flows of rivers, streams and drains in order to produce designs that would withstand the effects of rain. If the climate was in what is now called a 'steady state', then they could rely entirely on the existing rainfall figures, and that is what they did – in fact the habit grew up of accepting thirty-year sets of data as sufficient to show reliable averages and extremes.

This constant climate was accepted as fact in the 1930s when Watson-Watt made his remark about getting above the weather – it is easier, after all, to imagine getting above something when you are confident you know its dimensions. Students of engineering in the 1960s were still learning that the climate did not change, and even in the early 1970s the steady state was widely assumed. By then, however, academic opinion was becoming uneasy. Of course people had known all along that the climate of the British Isles had been very different in the distant geological past, but those were time frames of 25,000 years at the very least and that was hardly a matter of concern when it came to planning the route of a motorway. Historians, however, could point to more recent anomalies. For example, only 2,000 years ago Hannibal's armies included elephants from his native North Africa, but the modern climate in that region could not possibly support elephants. Closer to home, the Domesday Book referred to thirty-eight vineyards in England in 1086, while only a handful existed in 1960. And then there were the famous 'frost fairs' of late Tudor and Stuart times, when the River Thames froze so hard that thousands of Londoners could congregate on the ice. All of these suggested that the climate did change, even over periods as short as a few hundred years.

As this idea was digested (and a great deal of other

evidence was found to support it) the question naturally arose of what the current climate was up to, and the answer was interesting. Temperatures had risen by about one third of a degree Celsius between 1900 and 1940, but then they declined and there were very cold winters in the 1940s and again in 1962–3. So persistent was this cold spell that by the end of the 1960s newspapers and magazines were speculating about a long-term cooling of the earth, but then once again the line on the graph changed direction and a couple of warm winters in the 1970s dispelled the anxieties. Temperatures might wobble a little, it seemed, but not enough to shake the fundamental soundness of the steady-state climate. And then along came the evidence of Charles Keeling.

Keeling, a Seattle-born scientist at the University of California in San Diego, is in many respects an heir to George James Symons, for he has dedicated most of his career to the precise measurement of a single pheno-menon: the amount of carbon dioxide in the air. Carbon dioxide is well known now to be a greenhouse gas, one of the family of gases in the atmosphere that help to trap some of the solar heat radiated back off the surface of the earth – in short, they keep the planet warm. Carbon dioxide, it is also well known, is a gas that we produce when we burn fossil fuels, as we have been doing at an increasing rate since the beginning of the Industrial Revolution – around 1800 in fact. About a cen-tury ago it was first suggested that a moment would come when we had produced enough carbon dioxide to begin warming the earth noticeably (this was thought to be a welcome prospect). Concern about this was expressed in the 1940s and 1950s, but global tempera-tures were then falling so it did not seem to matter. Then in the mid-1950s Charles Keeling saw, as Symons had seen before him, that what was needed in a debate of this kind was hard evidence, so he set about measuring the amount of carbon dioxide in the air. To his surprise he quickly found that wherever he went the level was roughly the same: about 315 parts per million. From this he was able to deduce – and he was right – that this gas was so dispersed in the atmosphere, so thoroughly stirred into the air by the winds, that it was at the same level pretty well everywhere. More than that, he had the calculations of an English engineer, George Callendar, to suggest that in 1900 the level had been about 290 parts per million, so it seemed to have risen appreciably. From here it was a short step to establishing a constant series of measurements and this Keeling did in 1958 at Mauna Loa Observatory in Hawaii, a place blown by some of the freshest winds in the world. The 'Keeling Curve' showing the readings at Mauna Loa is now the most famous graph in the world. It is not a curve at all, because from 1958 the line moved steadily upwards.

If it was correct that carbon dioxide levels were rising and if it was also correct that this was likely to provoke a warming of the global climate, then temperatures should have been rising, and from the early 1970s that is precisely what they did. By the mid-1980s, with fifteen years of rising temperatures and twenty-five years of Keeling Curve, scientific opinion was becoming alarmed. By general consent a milestone was passed in 1988 when two things happened. First, James Hansen, a leading climatologist and senior figure at NASA, told a US Senate committee: 'The greenhouse effect has been detected and it is changing our climate now.' And second, the Intergovernmental Panel on Climate Change (IPCC), a body that came to involve thousands of scientists in dozens of countries, was established to investigate the problem. The IPCC has since produced three enormous reports and the conclusions of each have been broadly the same: the planet is warming and the balance of opinion suggests that carbon dioxide and other gases generated by human beings are contributing to this process and likely to accelerate it considerably in the coming century. A clear majority of scientists in relevant disciplines shares this view.

Over the whole of the twentieth century, even allowing for the mid-century dip, average temperatures around the world rose by just short of one degree Celsius. This may not seem much until you consider that it only took a shift of four degrees in the other direction to precipitate the last Ice Age. Climate modellers have been busy calculating the implications, both if the warming continues at its present rate and if it accelerates, as it might be expected to since our emissions of greenhouse gases are increasing rapidly. Their findings suggest dramatic changes for British rainfall. As the century progresses, they say, there will be more rain, not least because higher temperatures mean more evaporation off the seas. The south-east of England is likely to see an increase of at least one per cent in average rainfall, with areas to the north and west having higher increases, of at least two or three per cent. Again this may not sound much – it is equivalent to a handful more rainy days each year – but remember that these are minima, and some models show increases of twice as much and more. The real threat, however, is not in the total rainfall but in the way it is delivered. No one is quite sure of the physics but the prevailing view is that higher temperatures will alter both the distribution of rain throughout the year and the kinds of showers that we have.

Summers will become drier, in England at least, which means that all the more rain will come in the other seasons, especially autumn and winter. And there will be fewer light showers and more heavy downpours – what the scientists call intense events. This combination, of wetter winters and more frequent downpours, is the usual recipe for floods.

A word of caution: despite the words of James Hansen very few scientists, if any, claim that all this is a certainty, or even that the case for sustained warming of the climate is fully proven. They do not know enough about the long-term behaviour of the global climate to predict with real confidence what will happen over the next five years, let alone the next century. There are doubters, a minority but some of them influential, who feel that the effects of carbon dioxide emissions are being exaggerated and that recent warming can be explained in other ways. But this is not the place to air those debates; what matters here is that the century-old conviction that the climate was in a steady state, that it was a known and in most respects manageable process – in Britain at least – has gone. The best computer-modelling guess, as described above, may be wide of the mark, but not even the most hardened sceptic would be so reckless now as to base a planning or engineering decision of any importance on the assumption that our rainfall patterns in the twenty-first century will resemble those of the twentieth. That is why the Lewes flood report does not say that the town's defences should have been built higher in the 1960s or should be built higher now, and that is why the British government is telling the two or three million other people around the country who are most at risk: 'We cannot prevent floods.' This is a new humility.

An important indicator of this change is to be found in what the engineers call 'return periods' – estimates of how frequently an event of a given magnitude is likely to occur. The Lewes flood of 1960, for example, was said to have a return period of 100 years, or 1:100, and the new defences were built on the basis that this was a risk to be taken seriously. Small-scale projects such as local drains may be built to handle events with a return period of just 1:5, while a very big development such as the Thames Barrier, designed to protect London from large sea-surges, was built for a return period of 1:1000. Such periods are calculated by statistical methods and computer models drawing on vast amounts of data, and they have long had a powerful influence on planning – if a return period is long and the cost is also high, for example, a project may not be considered worth pursuing. These methods have helped to prevent a great deal of flood damage over the years, but in the light of fears about global warming and the end of the steady-state climate no one is sure how to calculate return periods any more.

People have undoubtedly been spooked by events such as the Lewes flood, where the peak flow was almost a metre (three feet) higher than the 1960 level, and the subsequent flood in York, where waters rose twenty-five centimetres (ten inches) above the highest levels seen in the twentieth century, but these are not in themselves statistically significant. You do not have to be an expert in probability theory to know that rare events sometimes happen. In the same way, the persistent wet weather we have experienced in Britain over recent years proves nothing, even though records have been broken; after all, most records are eventually broken. What makes a difference now is this new dimension of uncertainty that has been added by the prospect of long-term temperature rise, because it means that even 200-year data series may not be enough to enable us to estimate what is the worst the new climate can throw at us. People are having a stab at it. One paper in 2000, based on flow models for the rivers Thames and Severn, suggested that peak flood levels by 2050 could be 15 to 20 per cent higher than they are now. At a conference on flood risk in London in November 2001 a speaker from the Met Office's Hadley Centre for Climate Prediction and Research described the latest regional climate modelling exercises and gave a shocking picture of their impact on risk measurement. In some cases, Catherine Senior explained, it looked as though events previously estimated to have a 1:40 return period might have to be revised to a 1:4 period. But she cautioned – as people always do in this field – 'There really is still a lot of uncertainty.' In fact uncertainty proved to be the theme of that conference, with some mathematically inclined speakers admitting in effect that what they were looking for now was not so much a way of estimating return periods as a way to start estimating how unreliable their estimates were.

In this context it is generally accepted that a whole new approach to flooding is required, and even government ministers now speak of 'holistic solutions'. When the Institution of Civil Engineers was asked to look into the matter it produced a report with the revealing title *Learning to live with rivers*. While it stops some way short of complete surrender, the report accepts that past practice has been flawed and that in future we must treat our rivers and our rain with more respect.

It is not just people on river banks who need to be worried, or even people in other low-lying districts such as Lower Farringdon; almost all urban drainage is vulnerable. These systems are not usually built to cope with even the known long-term maximum flows, because that would be uneconomical; they are big enough only to handle the sort of flow that comes along on average once every five years or so, and older systems can usually handle less. They have the virtue, however (or at least it was seen as a virtue in the past), that they are quick, so that rainwater is rapidly swept off into progressively bigger drains and then into rivers which carry it to the sea. As our cities spread in area, and as they become more watertight, these systems have to be extended and upgraded to cope with more water. The emphasis is thus on *conveyance* – whisking the water away. What will happen to these systems in the event of frequent extreme rain events? The risk is that they will become like accelerated versions of river floods. The rain will fall by the millions of tons over large areas, landing on car parks, pavements, streets and roofs from which it will run directly or by gutter into the drains. If this proves too much for the drains that will mean ponds in the street and flooded cellars, often in districts which do not now consider themselves at risk. But even if the drains do cope the effect may well be damaging down the line, for in a matter of minutes a huge volume of water will flush through the system and out to the exit point, which in most cases will be a river already swelled by heavy winter rain. Efficient conveyance, therefore, creates new dangers, and it seems that the problem does not stop at our city boundaries. Research now indicates that one effect of modern intensive agriculture is to increase and accelerate the run-off of rain from farmland, so that even in the countryside the rivers are filling more quickly.

As less and less of the rain that falls on Britain is seeping into the ground, more and more of it is being hustled into our river systems. The traditional response has been to build ever-higher flood defences along the river banks, but with the new uncertainty of return periods it is difficult to know how high is high enough – and even if you do know, it may not be practical to create such structures. If you get it wrong, as in Lewes in 2000, you can actually make the flood worse. And if you get it right, successful defences usually contribute to the problem of conveyance. The flood is passed downstream so that each town on the route to the sea needs bigger defences to cope with faster-moving water and higher peak flows. Rivers are becoming giant open flush pipes.

Seeing the dangers, many engineers have been switching their attention from conveyance to *storage*. This does not just mean reservoirs, although they have their part to play. It means finding ways to slow the passage of the rainwater at any point where it can be held up. More houses, for example, should have water butts and these should be kept empty to be useful in a flood. Pavements and other surfaces should be porous, so that rainwater drains into the soil, the simplest form of storage and one of the most efficient. Open areas like playing fields and supermarket car parks should be adapted so that, if needed, they can be flooded with water that would otherwise overcharge the storm drains or the rivers. In the same way, large underground car parks could have a secondary use as emergency reservoirs. On a bigger scale, there should be 'managed retreat' from important floodplains, removing dykes and other barriers so that they can once again perform their function. This will not normally mean knocking down buildings (although it is certain that planners will approve far fewer new buildings in these areas) but rather the compulsory purchase of reclaimed farmland. The biggest step of all would be to plant more forests, because nothing soaks up water quite like trees – and they have the additional virtue of consuming carbon dioxide. Some of these things will happen, although it could be many years before they make a difference, and, as the Lewes flood shows, they are needed whether or not the warnings of the climate modellers prove justified. Even then it will be true, as the Environment Agency's mantra has it, that 'We cannot prevent floods. We can only prepare for them.'

This is a worrying world, very different from the one we knew a generation ago. It would be unfair to pick on Robert Watson-Watt, who was a scientist of distinction, as a symbol of misplaced confidence or pride. He wrote the quotation with which this narrative began at a time of steady-state climate and rapidly advancing knowledge, and so it must have seemed quite reasonable to him to think in terms of mastery of the weather. When he looked at civilization as he defined it, progress had been unrelenting for 150 years and the lives of British people had improved beyond measure. He had every reason for confidence; many others felt the same way and continued to do so for decades afterwards. A similar confidence, which now looks like complacency, led the Lewes fire service to build its headquarters where it could be flooded and the developers of Lower Farringdon to build the houses in Chase Field, and it was a common attitude. It turns out that a remarkable 1,800 hospitals and surgeries have been built in Britain on land that is vulnerable to flooding, and it was only recently that people began to accept this was a bad idea.

Now that we are not so confident, at the very least we can stop doing that.

If scientists, engineers and politicians have been slow to change their views, the public is likely to be slower still. Public expectations in these matters are high and rain usually seems a matter of little consequence, yet there must now be a managed retreat from the assumptions that science has the answers, that even if the price is high we can always buy protection, that we can cope with downpours and their consequences. I for one will have to get used to the idea that a soggy park is a useful thing, providing water storage and relieving pressure on the London drains. Many steps may be taken to change attitudes, and one of the simplest is a small, symbolic one. It is done today in parts of Germany and it used to be done in London before the Thames Barrier convinced everybody they were no longer at risk. Around every lamp post and every telegraph pole we could mark in bright yellow paint the peak level of the worst recorded flood in that area. That might remind us that, while at times we may get above ourselves, there is no such thing as getting above the rain.

cumulonimbus

Acknowledgements

In the course of my research I did my best to live up to Dr Johnson's dictum, raising the subjects of rain and weather with almost everyone I met. Most people indulged me, many offering useful thoughts and advice, and I would like to express my gratitude to them. In particular, I should thank the following: Annalisa Barbieri, Clive and Donna Butcher, Tim Cherrington, Colin Donnelly, Kate Fox, Lindsay Frost, Louise Goddard-Jones, Anne Hardy, Jacinta Kelly, Graham McLelland, Colin Maplesden, Alastair Muir, David Parker, David Sellers, Wendy Sykes and Tom Wilkie. All mistakes, of course, are my own and not theirs. I am also grateful to the staffs of the British Library and the libraries of the Wellcome Institute, the Royal Meteorological Society and the Royal Society. It was particularly useful to my research to attend the Royal Society's November 2001 conference on flood risk.

I found the following books most helpful. On general meteorology, these two are accessible and clear: *Teach Yourself: Weather*, by Ralph Hardy, 1996, and *How Weather Works: Understanding the Elements*, by René Chabaud, 1996. On British climatology the best general work is *Climates of the British Isles: Present, Past and Future*, edited by Mike Hulme and Elaine Barrow, 1997. I am conscious that I have concentrated on the British experience, but the wider picture can easily be found, notably in the works of H. H. Lamb, e.g. *Climate, History and the Modern World*, 1982. I also made use of *A History of the Theories of Rain*, by W. E. Knowles Middleton, 1965 and *Greenhouse: The 200-Year Story of Global Warming*, by Gale E. Christianson, 2000. Of the quoted works not identified in the text, Burney's remarks about Johnson appear in Boswell's *The Life of Samuel Johnson* as a foot-note for September 1784; Jeremy Paxman's book is *The English, A Portrait of a People*, published in 1998; the Farringdon local history is *The Last Hundred Years*, produced by the Farringdon and Chawton Magazine in 2000, and the details of the Salisbury experiments come from *The Common Cold*, Sir Christopher Andrewes's jaunty account of the work of his unit there, published in 1965. George James Symons has no biography and the information about him comes from his own writings, from obituary notices and from 'The British Rainfall Organization After Fifty Years', by Hugh Robert Mill, which appeared in *British Rainfall 1909*. A summary of the Environment Agency's excellent *Sussex Ouse 12th October 2000 Flood Report* can be read on the Lewes District Council website, www.lewes.gov.uk. The Meteorological Office website is www.metoffice.com.

Permissions